The Numerical Solution of the American Option Pricing Problem

Finite Difference and Transform Approaches

The Numerical Solution of the American Option Pricing Problem

Finite Difference and Transform Approaches

Carl Chiarella (University of Technology, Sydney, Australia)
Boda Kang (University of York, UK)
Gunter H. Meyer (Georgia Institute of Technology, USA)

World Scientific
NEW JERSEY • LONDON • SINGAPORE • BEIJING • SHANGHAI • HONG KONG • TAIPEI • CHENNAI

Published by

World Scientific Publishing Co. Pte. Ltd.
5 Toh Tuck Link, Singapore 596224
USA office: 27 Warren Street, Suite 401-402, Hackensack, NJ 07601
UK office: 57 Shelton Street, Covent Garden, London WC2H 9HE

Library of Congress Cataloging-in-Publication Data
Chiarella, Carl.
The numerical solution of the American option pricing problem : finite difference and transform approaches / Carl Chiarella (University of Technology, Sydney, Australia), Boda Kang (University of York, UK), Gunter H Meyer (Georgia Institute of Technology, USA).
pages cm
Includes bibliographical references and index.
ISBN 978-9814452618 (hardcover : alk. paper)
1. Options (Finance)--United States. 2. Options (Finance)--Mathematical models. I. Kang, Boda. II. Meyer, Gunter H. III. Title.
HG6024.U6C443 2014
332.64'23--dc23

2014021380

British Library Cataloguing-in-Publication Data
A catalogue record for this book is available from the British Library.

In-house Editors: Rajni Gamage/Chye Shu Wen

Typeset by Stallion Press
Email: enquiries@stallionpress.com

Printed in Singapore

Preface

This book is the outgrowth of two visits by Gunter Meyer to University of Technology, Sydney. It is also fortuitous that I had working with me, for over six years, Boda Kang. I took advantage of the fact that Boda was working with me and of Gunter's visits to complete this book. Basically it contains the output of a lot of work that the authors have undertaken, partly with a number of other co-authors who have been mentioned at various points in the text.

Carl Chiarella
August 2014

Contents

Chapter 1

Introduction

The American option pricing problem has been explored in great depth in the option pricing literature. The survey by Barone-Adesi (2005) provides an overview of this research for the case of the American put under the classical Brownian motion process for asset returns.

The literature probably started with the paper by McKean (1965) which was proposed to him by Samuelson (1965). Interestingly this paper was published about six years before the famous Black and Scholes (1973) paper. The first papers to give a numerical treatment probably was by Parkinson (1977) and Brennan and Schwartz (1977, 1978). Whilst the paper by Cox *et al.* (1979) on the binomial model provided a lot of intuitive appeal, it nevertheless took a long time to compute values accurately. Papers by Johnson (1983), Geske and Johnson (1984), MacMillan (1986), Barone-Adesi and Whaley (1987) and Ju and Zhong (1999) developed various approximate methods. One should also mention the work of Kim (1990), Jacka (1991a), Jamshidian (1992) and Carr *et al.* (1992) who provided put values as a function of the free boundary. Longstaff and Schwartz (2001) adapted Monte Carlo simulation methods to deal with the American put problem. The method gives the approximation of the free boundary by using a polynomial approximation to the continuation value. Monte Carlo methods have come a long way since the Longstaff and Schwartz paper but due to space limitations we shall not give details of these developments here as the focus of our book is more on partial differential equation methods.

We shall focus first on the formulation of the problem. In order to capture the market reality that the underlying asset returns are not well modelled by Brownian motion, it is necessary to consider both stochastic volatility and jump-diffusion dynamics as these are the processes most widely used to account for this effect. More recently, there has been a great deal of work on the use of Lévy processes, which is nicely surveyed in the book by Cont and Tankov (2003).

In this book we shall survey the integral transform approach, which in higher dimensions can only work for specific correlation structures, and the partial differential equation approach. For the latter we shall examine a number of finite difference approaches. Projected successive overrelaxation (PSOR), which when used with a high level of discretisation, will yield a benchmark solution against which we can compare other methods, such as the method of lines (MOL) for low dimensional problems and operator splitting methods for higher dimensional problems.

An array of approaches have been proposed in the vast literature on this topic. It will not be possible in this book to survey all of these. Rather we shall focus on the methods that the authors and some of their co-workers have found the most useful.

Chapter 2

The Merton and Heston Model for a Call

2.1 The Model

Let $C^A(S, v, \tau)$ be the price of an American call option written on a stock of price S with time to expiry τ[1] and strike price K. For the underlying dynamics, we assume that the stochastic differential equation (SDE) for S is given by the jump-diffusion process proposed by Merton (1976), in conjunction with the square root variance process of Heston (1993). Thus the dynamics for S under the so-called historical measure $\mathbb{P}$ are governed by the SDE system

$$dS = (\mu - \lambda k)Sdt + \sqrt{v}SdZ_1 + (Y-1)SdN, \tag{2.1}$$

$$dv = \kappa_v(\theta - v)dt + \sigma_v\sqrt{v}dZ_2. \tag{2.2}$$

In (2.1), μ is the instantaneous stock return per unit time, v is the instantaneous squared stock variance per unit time, and Z_1 is a standard Wiener process under $\mathbb{P}$. Furthermore, we define the Poisson jump arrival process N by

$$dN = \begin{cases} 1, & \text{with probability} \quad \lambda dt, \\ 0, & \text{with probability} \quad (1 - \lambda dt), \end{cases}$$

and set

$$k = \mathbb{E}_{\mathbb{P}}[(Y-1)] = \int_0^\infty (Y-1)G(Y)dY, \tag{2.3}$$

where $G(Y)$ is the continuous probability density function for the multiplicative jump sizes, Y, generated by the measure $\mathbb{P}$. In (2.2), θ is the long-run mean for v, κ_v is the rate of mean reversion, σ_v is the instantaneous volatility of v per unit time (the so-called vol-of-vol), and Z_2 is a standard

[1] Note that $\tau = T - t$, where T is the maturity date of the option and t is current time.

Wiener process under $\mathbb{P}$ correlated with Z_1 such that $\mathbb{E}[dZ_1 dZ_2] = \rho dt$. Note that dN, Y, dZ_1 and dZ_2 are otherwise uncorrelated and with this formulation we are not assuming any jump in the variance term, only in the stock price dynamics.

It is more convenient to convert the correlated Wiener processes to independent Wiener processes. This is accomplished by applying the Cholesky decomposition to the Wiener processes, Z_1 and Z_2 to obtain independent Wiener processes W_1 and W_2 such that,

$$\begin{aligned} dZ_1 &= dW_1, \\ dZ_2 &= \rho dW_1 + \sqrt{1-\rho^2} dW_2. \end{aligned} \tag{2.4}$$

Applying this transformation to equations (2.1) and (2.2) we obtain,

$$dS = (\mu - \lambda k) S dt + \sqrt{v} S dW_1 + (Y-1) S dN, \tag{2.5}$$

$$dv = \kappa_v(\theta - v)dt + \rho\sigma_v\sqrt{v}dW_1 + \sigma_v\sqrt{1-\rho^2}\sqrt{v}dW_2. \tag{2.6}$$

Now let r be the risk free rate of interest and assume that the underlying asset pays a continuously compounded dividend yield at a rate q (both r and q are assumed to be constant). Equations (2.5) and (2.6) are expressed under the real world probability measure, $\mathbb{P}$. However, for fair valuation of derivative securities, we need to work in the risk neutral world whose measure we denote as $\mathbb{Q}$. The transformation from $\mathbb{P}$ to $\mathbb{Q}$ is accomplished by applying Girsanov's Theorem for Wiener processes, application of which to equations (2.5) and (2.6) gives,

$$dS = (r - q - \lambda k) S dt + \sqrt{v} S d\tilde{W}_1 + (Y-1) S dN, \tag{2.7}$$

$$dv = \kappa_v(\theta - v)dt - \lambda(S,v,t)\sigma_v\sqrt{v}dt + \rho\sigma_v\sqrt{v}d\tilde{W}_1 \tag{2.8}$$

$$+\sigma_v\sqrt{v}\sqrt{1-\rho^2}d\tilde{W}_2, \tag{2.9}$$

where $\lambda(t,S,v)$ is the market price of volatility risk, and $\tilde{W}_1$ and $\tilde{W}_2$ are Wiener Processes under the measure $\mathbb{Q}$. The market price of volatility risk must be strictly positive as investors expect a positive risk premium in order for them to hold a risky security.

In determining the market prices of the volatility risk, we follow the arguments of Heston (1993) and assume that, $\lambda(S,v,t) = \frac{\lambda\sqrt{v}}{\sigma_v}$, where λ is a constant. Substituting this into equation (2.8) we obtain,

$$dS = (r - q - \lambda k) S dt + \sqrt{v} S d\tilde{W}_1 + (Y-1) S dN, \tag{2.10}$$

$$dv = [\kappa_v\theta - (\kappa_v + \lambda)v]dt + \rho\sigma_v\sqrt{v}d\tilde{W}_1 + \sqrt{1-\rho^2}\sigma_v\sqrt{v}d\tilde{W}_2. \tag{2.11}$$

Using standard hedging arguments and an application of Ito's lemma for jump-diffusion processes, it can be shown[2] that C^A satisfies the partial-integro differential equation (PIDE)

$$\begin{aligned}\frac{\partial C^A}{\partial \tau} =& \frac{vS^2}{2}\frac{\partial^2 C^A}{\partial S^2} + \rho\sigma_v v S\frac{\partial^2 C^A}{\partial S \partial v} + \frac{\sigma_v^2 v}{2}\frac{\partial^2 C^A}{\partial v^2} \\ &+ \left(r - q - \lambda\int_0^\infty (1-\lambda_J(Y))(Y-1)G(Y)dY\right) S\frac{\partial C^A}{\partial S} \\ &+ (\kappa_v\theta - (\kappa_v + \lambda_v)v)\frac{\partial C^A}{\partial v} \\ &- rC + \lambda\int_0^\infty (1-\lambda_J(Y))[C^A(SY,v,\tau) - C^A(S,v,\tau)]G(Y)dY,\end{aligned} \tag{2.12}$$

in the region $0 \le \tau \le T, \quad 0 < S \le a(v,\tau)$, and $0 \le v < \infty$,

$$C^A(S,v,\tau) = S - K, \ a(v,\tau) < S < \infty. \tag{2.13}$$

Here $a(v,\tau)$ denotes the early exercise boundary at time to maturity τ and variance level v, and $\lambda_J(Y)$ denotes the market price of risk associated with a jump in the value of the stock with magnitude Y (that is a jump from S to SY). At $S = a(v,\tau)$ we have the value matching condition

$$C^A(a(v,\tau),v,\tau) = a(v,\tau) - K, \tag{2.14}$$

and to avoid arbitrage opportunities,

$$\lim_{S\to a(v,\tau)} \frac{\partial C^A}{\partial S} = 1, \qquad \lim_{S\to a(v,\tau)} \frac{\partial C^A}{\partial v} = 0. \tag{2.15}$$

The boundary conditions (2.15) are referred to in the literature as the smooth-pasting conditions, and these follow by assuming that option holders will select their exercise strategy so as to maximise the value of the American call option. Mathematically, this is equivalent to ensuring that $\partial C^A/\partial S$ and $\partial C^A/\partial v$ will be continuous for all values of S. Figure 2.1 demonstrates the payoff, price profile and early exercise boundary for the American call under consideration.

[2]See for example Cheang *et al.* (2013). The paper also discusses the role that the assumptions concerning the market prices of volatility risk and jump risk play in choosing a risk-neutral pricing measure $\mathbb{Q}$.

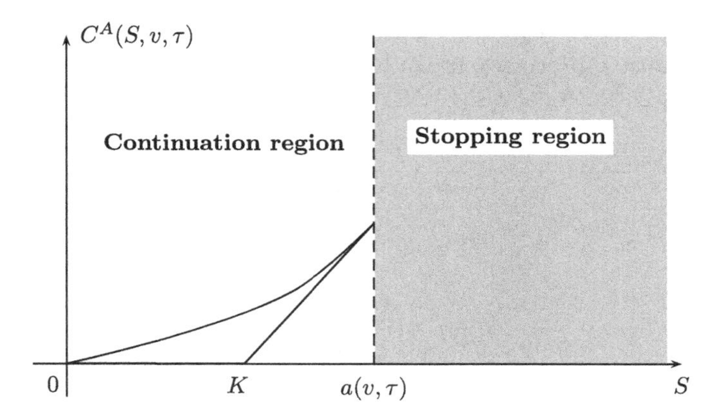

Fig. 2.1 Continuation and stopping regions for the American call option at a given value of v.

The PIDE (2.12) may be rewritten in the form

$$\frac{\partial C^A}{\partial \tau} = \frac{vS^2}{2}\frac{\partial^2 C^A}{\partial S^2} + \rho\sigma_v vS\frac{\partial^2 C^A}{\partial S\partial v} + \frac{\sigma_v^2 v}{2}\frac{\partial^2 C^A}{\partial v^2} + (r - q - \lambda^* k^*)S\frac{\partial C^A}{\partial S}$$
$$+ (\alpha - \beta v)\frac{\partial C^A}{\partial v} - (r + \lambda^*)C^A + \lambda^* \int_0^\infty C^A(SY, v, \tau)G^*(Y)dY, \tag{2.16}$$

where

$$\lambda^*\mathbb{E}_{\mathbb{Q}}[F(Y)] = \lambda^* \int_0^\infty F(Y)G^*(Y)dY$$
$$\equiv \lambda \int_0^\infty F(Y)(1 - \lambda_J(Y))G(Y)dY, \tag{2.17}$$

and

$$k^* \equiv \mathbb{E}_{\mathbb{Q}}[(Y - 1)], \tag{2.18}$$

with $\alpha \equiv \kappa_v\theta$ and $\beta \equiv \kappa_v + \lambda_v$. We note that

$$\int_0^\infty C^A(S, v, \tau)G^*(Y)dY = C^A(S, v, \tau),$$

where G^* is a density function under the risk-neutral pricing measure $\mathbb{Q}$.

The initial condition for (2.12) is the American call payoff function, given by

$$C^A(S, v, 0) \equiv c(S, v) = (S - K)^+ = \max\{S - K, 0\}. \tag{2.19}$$

To complete the mathematical model for the SVJD call (2.16) additional boundary conditions are required, but boundary conditions at $S = 0, v = 0$

and $v = v_{\max} \gg 0$ cannot be set solely by financial considerations because the structure of the equation determines what boundary conditions are admissible.

We do not know for what boundary conditions one can prove that the PIDE (2.16) for the call has a smooth solution, that is, under what conditions it is "well posed". But if we *assume* that the PIDE has a smooth solution, then we can in principle evaluate the jump integral and the derivative $\partial C^A/\partial \tau$ at this solution so that both terms can be considered a source term in (2.16). For any value of τ this makes (2.16) a second order elliptic partial differential equation in the $S - v$ plane for which there exists some information on admissible boundary data.

The coefficient matrix associated with (2.16) is

$$A(S,v) = \begin{pmatrix} \frac{vS^2}{2} & \frac{\rho\sigma_v vS}{2} \\ \frac{\rho\sigma_v vS}{2} & \frac{\sigma_v^2 v}{2} \end{pmatrix} \tag{2.20}$$

For $\rho^2 < 1$ this matrix is singular at $S = 0$ and $v = 0$. We can now apply the theory for second order equations with non-negative characteristic form of Oleinik and Radkevic (1973) to determine those boundaries where no boundary conditions can be imposed on (2.16).

On $S = 0$ the inward unit normal into the computational domain $S, v > 0$ (this means that S and v are both greater than zero) is $\vec{n} = (n_1, n_2) = (1, 0)$, and on $v = 0$ we have $\vec{n} = (0, 1)$. In either case we have the dot product

$$\langle A(S,v)\vec{n}, \vec{n}\rangle = 0.$$

It is known from Oleinik and Radkevic (1973) that admissible boundary conditions are related to the algebraic sign of the so-called Fichera function associated with $A(S,v)$. The Fichera function, determined by the elliptic part of (2.16), is

$$\begin{aligned} h(S,v) = &\left[(r - q - \lambda^* k^*)S - \left(vS + \frac{\rho\sigma_v S}{2}\right)\right] n_1 \\ &+ \left[(\alpha - \beta v) - \left(\frac{\rho\sigma_v v}{2} + \frac{\sigma_v^2}{2}\right)\right] n_2. \end{aligned}$$

We see that

$$h(0,v) = 0$$

because $\vec{n} = (1, 0)$. It now follows from Oleinik and Radkevic (1973) that on $S = 0$ no boundary condition should be imposed, and if we assume that the postulated solution of the PIDE (2.16) has bounded derivatives, then the equation (2.16) must hold. Hence we have the boundary condition on $S = 0$:

$$\frac{\sigma_v^2 v}{2}\frac{\partial^2 C^A}{\partial v^2} + (\alpha - \beta v)\frac{\partial C^A}{\partial v} - rC^A - \frac{\partial C^A}{\partial \tau} = 0, \; 0 < v < \infty. \tag{2.21}$$

It is common to use on $S = 0$ the boundary value

$$C^A(0, v, \tau) = 0. \tag{2.22}$$

It is implied by the stochastic differential equation (2.1) because if S reaches 0 it stays zero and the call is worthless. We note that (2.22) is a solution of (2.21) provided one does not impose an inconsistent boundary condition at a computational right boundary $v = v_{\max}$. In the literature the boundary condition

$$\frac{\partial C(S, v_{\max}, \tau)}{\partial v} = 0 \tag{2.23}$$

is frequently chosen for numerical simulations. It is consistent with (2.22) and implies that the option becomes insensitive to the volatility as $v \to \infty$.

On the boundary $v = 0$ we find that

$$h(S, 0) = \alpha - \frac{\sigma_v^2}{2}.$$

If $h(S, 0) \geq 0$ which is Feller condition, then according to Oleinik and Radkevic (1973) the PIDE (2.16) with $v = 0$ should hold, which takes on the form

$$(r - q - \lambda^* k^*)S\frac{\partial C^A}{\partial S} + \alpha\frac{\partial C^A}{\partial v} - (r + \lambda^*)C^A$$
$$+ \lambda^* \int_0^\infty C^A(SY, 0, \tau)G^*(Y)dY - \frac{\partial C^A}{\partial \tau} = 0. \tag{2.24}$$

If $h(S, 0) < 0$ then other boundary conditions, for example Dirichlet conditions, could be imposed.

If the integral is again considered just a source term, then (2.24) can also be interpreted as an admissible oblique boundary condition for the inhomogeneous parabolic equation (2.16) because the convective terms of (2.16) point inward on $v = 0$, so that

$$\langle((r - q - \lambda^* k^*)S, \alpha), (0, 1)\rangle = \alpha > 0, \tag{2.25}$$

where $< \overrightarrow{x}, \overrightarrow{y} >$ denotes the dot product of $\overrightarrow{x}$ and $\overrightarrow{y}$.

Such boundary conditions are a generalization of the Neumann and Robin boundary condition and lead to well posed problems for elliptic and parabolic partial differential equations (see Lieberman (1996)).

There is no guarantee at this time that the PIDE (2.16) with the boundary conditions (2.21) or (2.22), (2.24), the free boundary conditions (2.13–2.15) and an additional condition on a computational boundary $v = v_{\max}$ (as discussed in Chapter 7) will lead to a well posed problem for the PIDE (2.16). However, if the problem is well posed then the solution should be continuous with respect to the financial parameters regardless of the algebraic sign of the Fichera function. Here (2.24), interpreted as an oblique boundary condition will always be the boundary condition for (2.16) on $v = 0$ regardless of the Feller condition.

For a more expansive discussion of boundary condition for (2.16) we refer to Meyer (2015). We also mention that first analytic results for the well posedness of a PIDE with oblique boundary conditions are beginning to appear in the mathematical literature (see for example, Barles *et al.* (2013)).

Option pricing under stochastic volatility and jump-diffusion dynamics of course involves market incompleteness since both the variance risk and the jump risk are not priced in the market. When the market is incomplete option pricing formulae are not unique. In (2.16) the non uniqueness is reflected in the market price of variance risk parameter λ_v (embedded in the parameter β) and the market price of jump risk $\lambda_J(Y)$ that is embedded in the parameter λ^* and jump-size distribution under the adjusted measure $\mathbb{Q}$ (see equation (2.17)). There is a large literature on how these parameters may be chosen, for instance by minimising the variance of hedging cost or some entropy measure. Here we shall simply assume that the parameters λ_v, λ^* as well as the parameters of the distribution G^* have somehow been obtained either by the methods referred to or simply by calibration to market data, which would probably be the method chosen by practitioners.

Our main task in Chapters 4 and 7 is to develop approaches to solving the PIDE (2.16) subject to the boundary conditions (2.15) and (2.19). As we indicated in the introduction we shall consider both the transform approach and numerical approaches. The main numerical challenges in solving equation (2.16) arise from, (i) the two spatial dimensions (S and v) so that we are seeking an early exercise surface, and (ii) the integral term over the jump-size distribution which will involve the unknown option price over a whole set of values of S.

Chapter 3

American Call Options under Jump-Diffusion Processes

In this chapter we shall drop the stochastic volatility component from the dynamics by assuming that the variance is constant and merely discuss how to handle the jump term in the transform approach. Option pricing under jump-diffusion dynamics was originally investigated by Merton (1976) for the case of the European option, but here we also consider American options.

3.1 Introduction

The integral transform technique can be applied to solve the option pricing problem under a range of underlying dynamics and payoff types. There are numerous examples where integral transforms have been used in the option pricing literature. Scott (1997) uses Fourier transforms to value European options under jump-diffusion with both stochastic volatility and stochastic interest rates. Carr and Madan (1999) with a particular focus on the variance gamma model use Fourier transforms to motivate numerical algorithms using the fast Fourier transform (FFT) technique. The American strangle has been analysed by Chiarella and Ziogas (2005) using a Fourier transform approach, under the assumption that the underlying dynamics are the classic geometric Brownian motion.

The main advantage of the Fourier transform method is that the solution may be expressed in terms of a general initial value or payoff function, so that a wider variety of American options such as puts, calls, butterflies, spread options and max-options can all be handled systematically.

In this chapter we derive the linked system of integral equations for the price and early exercise boundary of an American call under Merton's jump-diffusion dynamics. We also derive the limit of the early exercise boundary at maturity, and discuss a numerical algorithm for solving the linked integral equation system. We do so by extending to the jump-diffusion situation the quadrature integration technique of Kallast and Kivinukk (2003). The algorithm used has the added advantage of generating values for the price, delta and early exercise boundary. The issue of the existence and uniqueness of the solution to the non-linear integral equation that arises in the solution of the free boundary value problem was raised by Myneni (1992) and finally settled by Peskir (2005) in the case of pure diffusion dynamics. It still remains to extend that analysis to the case of jump-diffusion dynamics.

We focus on deriving solutions to the partial differential equation (PDE) for the American call price. A natural solution technique here is the integral transform approach, which (see Debnath (1995)) has been very successfully employed in a wide range of PDE problems in the natural sciences. One of the difficulties with dealing with American options using this technique arises from the fact that one has to solve the PDE on a region restricted by the early exercise, or free, boundary. McKean (1965), who seems to have been the first to consider the American option pricing problem, solves the homogeneous PDE in a restricted domain by using an incomplete Fourier transform. An alternative approach, developed by Jamshidian (1992), replaces the homogeneous PDE with an equivalent inhomoegenous PDE on an unrestricted domain. The solution to this alternative formulation can then be derived by using the standard Fourier transform, and then through an application of Duhamel's principle. In this chapter we provide this extension of Jamshidian's formulation to the jump-diffusion case.

While some authors, in particular Pham (1997) and Gukhal (2001), derive integral equations for the price and free boundary of American options under jump-diffusion they do not discuss how they can be solved numerically. In this section we solve the linked integral equation system that arises for the American call and its free boundary in the case of jump-diffusion dynamics. In developing a numerical scheme for the linked integral equation system we obtain a simplification of the integral terms over the jump-size distribution that reduces the computational burden by reducing the dimension of the multiple integrals involved. We give some results on the behaviour of the early exercise boundary at expiry that are needed to start the numerical procedure. We use the proposed numerical integration scheme to find the price, delta and early exercise boundary of American calls with log-normal jump sizes. We also demonstrate that the method is often more efficient than a simple two-pass Crank-Nicolson finite difference scheme and is competitive with the method of lines of Meyer (2015, 1998).

Sec. 3.2 outlines the free boundary problem that arises from pricing an American call option under Merton's jump-diffusion model. Sec. 3.3 applies Jamshidian's method to derive an inhomogeneous PIDE for the American call price, which is then solved using Fourier transforms. We thus obtain the linked integral equation system for the free boundary and option price. We derive the limit of the free boundary at expiry in Sec. 3.4, and find that this limit is different to that found for pure diffusion models. Sec. 3.5 analyses the integral equations in the case where the jump sizes follow a log-normal distribution, as suggested by Merton (1976). Sec. 3.6 gives properties of the free boundary at expiry. Sec. 3.7 outlines the numerical integration method used to solve the linked integral equation system for both the free boundary, price and delta of the American call. Since the integral equation for the call value and the integral equation for the free boundary are interdependent, we provide a method to handle this interdependence in order to use a two-pass sequential procedure that works well in the non-jump case. Numerical results detailing the efficiency of this algorithm are provided in Sec. 3.8.

3.2 The Problem Statement — Merton's Model

As we drop the stochastic volatility component of Chapter 2 we simply denote the price of an American option written on the underlying asset S at time to expiry τ by $C^A(S,\tau)$. In this case the free boundary only depends only on τ so we write it as $S = a(\tau)$. Because in this section $\sigma = \sqrt{v}$ is constant, the dynamics for S follows the jump-diffusion process (2.5), where in addition to the notation introduced in Chapter 2 we set the instantaneous variance per unit time as a constant σ and W_1 (becomes W because there is no stochastic volatility term) is a standard Wiener process under the physical measure.

The PIDE equation (2.12) for C^A in the current case becomes[1]

$$\frac{\partial C^A}{\partial \tau} = \frac{1}{2}\sigma^2 S^2 \frac{\partial^2 C^A}{\partial S^2}$$

[1]Pham (1997), to the authors' knowledge, was the first to report the PIDE (3.1) with the inclusion of the $\lambda_J(Y)$ term.

$$+\left(r-q-\lambda\int_0^\infty (Y-1)(1-\lambda_J(Y))G(Y)dY\right)S\frac{\partial C^A}{\partial S}-rC^A$$

$$+\lambda\int_0^\infty [C^A(SY,\tau)-C^A(S,\tau)](1-\lambda_J(Y))G(Y)dY, \qquad (3.1)$$

which has to be solved in the region $0 \le \tau \le T$ and $0 \le S \le a(\tau)$. Given a form for $\lambda_J(Y)$, we can define a new intensity, λ^*, and jump-size distribution, $G^*(Y)$, which fully incorporate the term $\lambda_J(Y)$, such that (3.1) can be written as

$$\frac{\partial C^A}{\partial \tau} = \frac{1}{2}\sigma^2 S^2 \frac{\partial^2 C^A}{\partial S^2}$$
$$+\left(r-q-\lambda^*\int_0^\infty (Y-1)G^*(Y)dY\right)S\frac{\partial C^A}{\partial S}-rC^A$$
$$+\lambda^*\int_0^\infty [C^A(SY,\tau)-C^A(S,\tau)]G^*(Y)dY. \qquad (3.2)$$

The market incompleteness of the jump-diffusion option pricing problem is reflected in the fact that the choice of λ^* and $G^*(Y)$ is at the discretion of the model builder, so that the associated risk-neutral density is non-unique. As stated earlier, we take the view here that the model builder would calibrate λ^* and $G^*(Y)$ directly to market data without needing to determine a risk-neutral density for the pricing problem at hand. Henceforth, for ease of notation, we shall simply refer to the jump risk-adjusted intensity and jump-size density in (3.2) as λ and $G(Y)$ respectively. Thus the expected jump size k, is given by

$$k = \mathbb{E}_Q[Y-1] = \int_0^\infty (Y-1)G(Y)dY.$$

By use of equation (2.3), the PIDE (3.2) can be written as

$$\frac{\partial C^A}{\partial \tau} = \frac{1}{2}\sigma^2 S^2 \frac{\partial^2 C^A}{\partial S^2} + (r-q-\lambda k)S\frac{\partial C^A}{\partial S}$$
$$-rC^A + \lambda\int_0^\infty [C^A(SY,\tau)-C^A(S,\tau)]G(Y)dY. \qquad (3.3)$$

Since we are considering American call options, the initial and boundary conditions equation (2.15) and equation (2.19) become

$$C^A(S,0) = (S-K)^+, \quad 0 \le S < \infty \qquad (3.4)$$
$$C^A(0,\tau) = 0, \quad \tau \ge 0, \qquad (3.5)$$

$$C^A(a(\tau), \tau) = a(\tau) - K, \quad \tau \geq 0, \tag{3.6}$$

$$\lim_{S \to a(\tau)} \frac{\partial C^A}{\partial S} = 1, \quad \tau \geq 0, \tag{3.7}$$

and the PIDE (3.3) is solved subject to these.[2]

If we were to use the approach of McKean (1965) we would introduce an incomplete version of the Fourier transform, defined by (note that $x = \ln(S)$, $V(x, \tau) = C^A(e^x, \tau)$ and $b(\tau) = a(\tau)/K$)

$$\mathcal{F}^b\{V(x, \tau)\} \equiv \int_{-\infty}^{\ln b(\tau)} e^{i\eta x} V(x, \tau) dx. \tag{3.8}$$

Inversion of the transform is readily carried out, and the end result is a linked system of integral equations for $C^A(S, \tau)$ and $a(\tau)$. This solution method comes at the expense of the drawbacks reported by Chiarella and Ziogas (2009) namely, the fact that the solution depends upon the derivative of the free boundary, which is difficult to calculate.[3] We instead seek a more efficient alternative for solving (3.3)–(3.7) using the standard Fourier transform.

3.3 Jamshidian's Representation

One of the major difficulties in solving the PIDE (3.3) subject to the boundary conditions (3.4)–(3.7) is that the solution is sought on a restricted domain for the stock price S, and furthermore the boundary of this domain is itself unknown a priori and needs to be determined as part of the solution process. In the pure diffusion case, Jamshidian (1992) demonstrates that by evaluating the PDE for the American call price when $S > a(\tau)$, one can reformulate the free boundary problem in the restricted domain $0 \leq S \leq a(\tau)$ as an inhomogeneous PDE in the unrestricted domain $0 \leq S < \infty$. This inhomogeneous PDE can then be more readily solved by traditional solution techniques such as Fourier transforms. We note that

[2] The smooth-pasting condition (3.7) sets the delta of the American call to be continuous at the free boundary so as to guarantee arbitrage-free prices. For the call under consideration, we note that the standard arbitrage arguments that justify condition (3.7) are not readily applied under Merton's jump-diffusion model, since this depends upon the price process for S being continuous. The corresponding boundary conditions were proven by Pham (1997) for the American put case, and the arguments he uses, based on the maximum principle, are readily applied to the case of the American call problem with a continuous dividend yield for S.

[3] Full details on McKean's method applied to American calls under jump-diffusion are given by Chiarella and Ziogas (2006).

the free boundary value problem (FBVP) given by (3.3)–(3.7) involves a homogeneous PIDE to be solved in the restricted asset price domain $0 \leq S \leq a(\tau)$. We shall apply Jamshidian's approach to reformulate the PIDE (3.3) and associated boundary conditions as an inhomogeneous PIDE on an unrestricted domain.

We highlight the fact that $C^A(S,\tau)$ and $\partial C^A/\partial S$ are continuous for $0 \leq S < \infty$ and $\tau > 0$, as given by the value-matching condition (3.6) and smooth-pasting condition (3.7). Jamshidian's approach can only be applied with confidence when such continuity holds. We now state the main result that converts the homogeneous PIDE on a restricted domain to an inhomogeneous PIDE on an unrestricted domain.

Proposition 3.1. *The solution to the homogeneous PIDE* (3.3) *for* $C^A(S,\tau)$ *in the domain* $0 \leq S \leq a(\tau)$ *subject to the initial and boundary conditions* (3.4)–(3.7) *is equivalent to the solution to the inhomogeneous PIDE*

$$\begin{aligned}\frac{\partial C^A}{\partial \tau} &= \frac{1}{2}\sigma^2 S^2 \frac{\partial^2 C^A}{\partial S^2} + (r - q - \lambda k) S \frac{\partial C^A}{\partial S} - rC^A \\ &\quad + \lambda \int_0^\infty [C^A(SY,\tau) - C^A(S,\tau)]G(Y)dY + \mathbb{1}_{\{S-a(\tau)\geq 0\}} \\ &\quad \times \left\{ qS - rK - \lambda \int_0^{a(\tau)/S} [C^A(SY,\tau) - (SY-K)]G(Y)dY \right\},\end{aligned} \tag{3.9}$$

in the region $0 < \tau \leq T$, $0 \leq S < \infty$, *subject to the initial condition* (3.4), *where* $\mathbb{1}_{\{x\geq 0\}}$ *is the indicator function defined as*

$$\mathbb{1}_{\{x\geq 0\}} = \begin{cases} 1, & x \geq 0, \\ 0, & x < 0. \end{cases} \tag{3.10}$$

Proof. Whenever S is in the stopping region, it is optimal to exercise the American call option, and hence the call option price is given by $C^A(S,\tau) = S - K$ for all $S \geq a(\tau)$. Although $C^A(S,\tau)$ only satisfies the PIDE (3.3) for $0 \leq S \leq a(\tau)$, we can introduce an inhomogeneous term in (3.3) such that C^A satisfies the PIDE for all $S \geq 0$. Jamshidian (1992) demonstrated that this was possible under pure-diffusion dynamics, and here we extend his result to the jump-diffusion case.

To derive the required inhomogeneous term, we evaluate (3.3) when $C^A(S,\tau) = S - K$. Thus we have

$$\begin{aligned}\Psi(S,\tau) &\equiv \mathbb{1}_{\{S-a(\tau)\geq 0\}}\left\{\frac{1}{2}\sigma^2 S^2 \frac{\partial^2 C^A}{\partial S^2} + (r-q-\lambda k)S\frac{\partial C^A}{\partial S} - rC - \frac{\partial C^A}{\partial \tau}\right.\\ &\quad \left. +\lambda\int_0^\infty [C^A(SY,\tau) - C^A(S,\tau)]G(Y)dY\right\}\\ &= \mathbb{1}_{\{S-a(\tau)\geq 0\}}\\ &\quad \left\{K(r+\lambda) - S(q+\lambda[k+1]) + \lambda\int_0^\infty C^A(SY,\tau)G(Y)dY\right\},\end{aligned}$$

where $\mathbb{1}$ is the indicator function given by (3.10). The indicator function is used to denote that Ψ is only valid for $S \geq a(\tau)$. Since $C^A(SY,\tau) = S - K$ when $SY \geq a(\tau)$, we can express $\Psi(S,\tau)$ as

$$\begin{aligned}\Psi(S,\tau) &= \mathbb{1}_{\{S-a(\tau)\geq 0\}}\left\{K(r+\lambda) - S(q+\lambda[k+1])\right.\\ &\quad \left. +\lambda\left(\int_0^{a(\tau)/S} C^A(SY,\tau)G(Y)dY + \int_{a(\tau)/S}^\infty (SY-K)G(Y)dY\right)\right\}\\ &= \mathbb{1}_{\{S-a(\tau)\geq 0\}}\left\{K(r+\lambda) - S(q+\lambda[k+1])\right.\\ &\quad +\lambda\int_0^{a(\tau)/S} C^A(SY,\tau)G(Y)dY\\ &\quad \left. +\lambda\left(\int_0^\infty (SY-K)G(Y)dY - \int_0^{a(\tau)/S}(SY-K)G(Y)dY\right)\right\}.\end{aligned}$$

Recalling that $k = \mathbb{E}^{Q_Y}[Y-1]$, we have

$$\begin{aligned}\Psi(S,\tau) &= \mathbb{1}_{\{S-a(\tau)\geq 0\}}\\ &\quad \times\left\{rK - qS + \lambda\int_0^{a(\tau)/S}[C^A(SY,\tau) - (SY-K)]G(Y)dY\right\}.\end{aligned} \tag{3.11}$$

Since (3.11) is the value of the PIDE (3.3) when $S \geq a(\tau)$, we can rewrite the PIDE as

$$\begin{aligned}\frac{\partial C^A}{\partial \tau} &= \frac{1}{2}\sigma^2 S^2 \frac{\partial^2 C^A}{\partial S^2} + (r - q - \lambda k) S \frac{\partial C^A}{\partial S} - rC \\ &\quad + \lambda \int_0^\infty [C^A(SY,\tau) - C^A(S,\tau)] G(Y) dY + \mathbb{1}_{\{S - a(\tau) \geq 0\}} \\ &\quad \times \left\{ qS - rK - \lambda \int_0^{a(\tau)/S} [C^A(SY,\tau) - (SY - K)] G(Y) dY \right\},\end{aligned}$$

which is equation (3.9) from Proposition 3.1. Note that it is easy to verify that (3.9) is satisfied by $C^A(S,\tau)$ for $0 \leq S < \infty$. □

Gukhal (2001) provides a clear economic interpretation for the inhomogeneous term that arises in equation (3.9). The $(qS - rK)$ term represents the net cash flows received from holding the portfolio $(S - K)$ whenever S is in the stopping region. This is already familiar from the pure diffusion case (see for example Kim (1990)). The integral term arises entirely because of the introduction of jumps in the price process for S. Note that if no jumps were present (λ=0) then this term would be zero, and the inhomogeneous term would become the same one presented by Jamshidian (1992). This additional term captures the rebalancing costs incurred by the option holder whenever the price of the underlying jumps down from the stopping region back into the continuation region. Figure 3.1 illustrates this effect in detail. Consider the case where, during the life of the option contract, the underlying asset price is at $S_- > a(\tau)$. Since the value of S is in the stopping region, the holder of the option will currently possess the portfolio $(S - K)$. If a jump of size Y occurs such that $S_+ = YS_- < a(\tau)$, then the portfolio held by the investor will now be worth less than the unexercised American call. This difference is the cost being captured by the integral in the inhomogeneous term in (3.9).

Having derived the inhomogeneous PIDE for $C^A(S,\tau)$ we next use Fourier transforms to find the solution. Our first step is to transform the PIDE to an equation with constant coefficients and a "standardised" strike of 1. Let $S \equiv Ke^x$ and $C^A(S,t) \equiv KV(x,\tau)$, with $b(\tau) \equiv a(\tau)/K$. The transformed PIDE for V is then

$$\frac{\partial V}{\partial \tau} = \frac{1}{2}\sigma^2 \frac{\partial^2 V}{\partial x^2} + \phi \frac{\partial V}{\partial x} - (r + \lambda) V$$

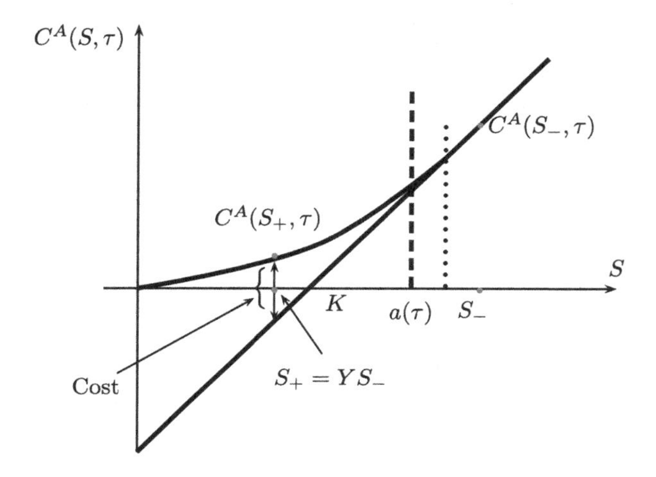

Fig. 3.1 Cost incurred by the investor from downward jumps in S.

$$+\lambda \int_0^\infty V(x + \ln Y, \tau)G(Y)dY + \mathbb{1}_{\{x - \ln b(\tau) \geq 0\}}$$
$$\times \left\{ \lambda \int_0^{b(\tau)e^{-x}} [V(x + \ln Y, \tau) - (Ye^x - 1)]G(Y)dY \right\}, \tag{3.12}$$

where $\phi \equiv r - q - \lambda k - \sigma^2/2$. Equation (3.12) is to be solved in the time domain $0 \leq \tau \leq T$, and the unrestricted region $-\infty \leq x \leq \infty$, subject to the initial and boundary conditions

$$V(x, 0) = (e^x - 1)^+, \quad -\infty < x < \infty, \tag{3.13}$$
$$\lim_{x \to -\infty} V(x, \tau) = 0, \quad \tau \geq 0, \tag{3.14}$$
$$V(\ln b(\tau), \tau) = b(\tau) - 1, \quad \tau \geq 0. \tag{3.15}$$

It is worth noting that the smooth-pasting condition still holds, although we do not explicitly require it when solving (3.12) for $V(x, \tau)$ since it is incorporated in the inhomogeneous term.[4]

Since the x-domain is now the unrestricted region $-\infty < x < \infty$, the Fourier transform of the inhomogeneous PIDE (3.12) can be found. Define

[4]We again refer the reader to the remark just prior to Proposition 3.1, in particular, the first sentence of that paragraph.

the Fourier transform of V, $\mathcal{F}\{V(x,\tau)\}$, as

$$\mathcal{F}\{V(x,\tau)\} \equiv \hat{V}(\eta,\tau) = \int_{-\infty}^{\infty} e^{i\eta x} V(x,\tau)dx, \tag{3.16}$$

with the corresponding inversion formula

$$\mathcal{F}^{-1}\{\hat{V}(\eta,\tau)\} = \frac{1}{2\pi}\int_{-\infty}^{\infty} e^{-i\eta x}\hat{V}(\eta,\tau)d\eta, \tag{3.17}$$

where i is the complex number. Applying this Fourier transform to (3.12), we can reduce the inhomogeneous PIDE for V to an inhomogeneous ordinary differential equation (ODE) for $\hat{V}$, whose solution is readily found.

When taking the Fourier transform of (3.12), we note that $V(x,\tau)$ and $\partial V(x,\tau)/\partial x$ do not approach zero as $x \to \infty$. A number of approaches have been suggested for dealing with this difficulty. Carr and Madan (1999) and Lee (2004) suggest introducing a damping function of the form $e^{-\alpha x}$ for some positive constant α, and then apply the transform to the dampened option price $U(x,\tau) = e^{-\alpha x}V(x,\tau)$, which does tend to zero, along with $\partial U(x,\tau)/\partial x$, as $x \to \infty$. The desired function $V(x,\tau)$ can be readily recovered after the solution in transform space has been inverted. Lewis (2000) gives another approach which proves that the Fourier transform is still valid when solving for $V(x,\tau)$ in this case, although one must instead take the complex Fourier transform in a strip of the complex plane. Regardless of which approach is used to make the Fourier transform applicable, it turns out that both methods are equivalent to simply assuming that $V(x,\tau)$ and $\partial V(x,\tau)/\partial x$ tend to zero as $x \to \infty$, and applying the standard transform accordingly. After this the solution may then be substituted into the PIDE (3.9) in order to verify that it satisfies this equation. Thus in order to simplify the technical discussion, we shall simply apply this assumption and suppress the finer details involved.[5]

Proposition 3.2. *Using the initial and boundary conditions* (3.13)–(3.14), *the Fourier transform of the PIDE* (3.12) *with respect to x satisfies the ODE*

$$\frac{\partial \hat{V}}{\partial \tau} + \left[\frac{\sigma^2\eta^2}{2} + \phi i\eta + (r+\lambda) - \lambda A(\eta)\right]\hat{V} = \hat{F}_J(\eta,\tau) \tag{3.18}$$

[5]Another approach, used by Cheang *et al.* (2013) involves applying the transform to the Kolmogorov PDE for the transition density (where the boundedness conditions hold more naturally) and then using Duhamel's principle to express the solution.

where

$$\hat{F}_J(\eta,\tau) \equiv \mathcal{F}\{F_J(x,\tau)\},$$

with

$$F_J(x,\tau) = \mathbb{1}_{\{S-a(\tau)\geq 0\}} \left\{(qe^x - r) - \lambda \int_0^{b(\tau)e^{-x}} [V(x+\ln Y,\tau) - (Ye^x - 1)]G(Y)dY\right\} \tag{3.19}$$

and

$$A(\eta) \equiv \int_0^\infty e^{-i\eta \ln Y} G(Y)dY. \tag{3.20}$$

Furthermore, the solution to the ODE (3.18) *is given by*

$$\hat{V}(\eta,\tau) = \hat{V}(\eta,0)\exp\left\{-\left(\frac{1}{2}\sigma^2\eta^2 + \phi i\eta + (r+\lambda) - \lambda A(\eta)\right)\tau\right\} + \int_0^\tau \exp\left\{-\left(\frac{1}{2}\sigma^2\eta^2 + \phi i\eta + (r+\lambda) - \lambda A(\eta)\right)(\tau-\xi)\right\} \times \hat{F}_J(\eta,\xi)d\xi, \tag{3.21}$$

where $\hat{V}(\eta,0) = \mathcal{F}\{V(x,0)\}$.

Proof. For the inhomogeneous term, we have $\hat{F}_J(\eta,\tau) \equiv \mathcal{F}\{F_J(x,\tau)\}$, and the only term that needs to be evaluated is the one involving the integral, namely

$$\mathcal{F}\left\{\int_0^\infty V(x+\ln Y,\tau)G(Y)dY\right\} = \int_{-\infty}^\infty e^{i\eta x}\int_0^\infty V(x+\ln Y,\tau)G(Y)dY dx. \tag{3.22}$$

Using the change of variable $z = x + \ln Y$, equation (3.22) becomes

$$\mathcal{F}\left\{\int_0^\infty V(x+\ln Y,\tau)G(Y)d(Y)\right\} = A(\eta)\hat{V}(\eta,\tau),$$

where $A(\eta)$ is defined in equation (3.20).

Hence, our PIDE is transformed into the ODE

$$\frac{\partial \hat{V}}{\partial \tau} + \left[\frac{\sigma^2\eta^2}{2} + \phi i\eta + (r+\lambda) - \lambda A(\eta)\right]\hat{V} = \hat{F}_J(\eta,\tau),$$

the solution of which is given by equation (3.21). □

The first term on the right-hand side in (3.21) is the Fourier transform of the PIDE (3.12) in the case of a European call option under jump-diffusion dynamics. In this case the last term of the right-hand side of (3.12) does not appear, and it is the latter term involving the free boundary that gives rise to the second term in equation (3.21).

Now that $\hat{V}(\eta, \tau)$ has been found, we may use the inversion formula (3.17) to recover $V(x, \tau)$, the American call price in the x-τ plane. By taking the inverse Fourier transform of (3.21), we have

$$\begin{aligned} V(x,\tau) &= \mathcal{F}^{-1}\Bigg\{\hat{V}(\eta,0)\exp\left\{-\left(\frac{1}{2}\sigma^2\eta^2+\phi i\eta+(r+\lambda)-\lambda A(\eta)\right)\tau\right\} \\ &\quad+\int_0^{\tau}\exp\left\{-\left(\frac{1}{2}\sigma^2\eta^2+\phi i\eta+(r+\lambda)-\lambda A(\eta)\right)(\tau-\xi)\right\} \\ &\quad\times\hat{F}_J(\eta,\xi)d\xi\Bigg\} \\ &\equiv V_E(x,\tau)+V_P(x,\tau) \\ &\equiv \frac{1}{K}\left[C_E(S,\tau)+C_P(S,\tau)\right]=\frac{1}{K}C^A(S,\tau), \end{aligned} \tag{3.23}$$

where $C_E(S,\tau) = KV_E(x,\tau)$ is the value of the corresponding European call written on S and $C_P(S,\tau) = KV_P(x,\tau)$ is the early exercise premium for $C^A(S,\tau)$. By performing the inversions, we can determine the analytic forms of C_E and C_P and these are given in propositions 3.3 and 3.4 below.

Proposition 3.3. *The price of the European call option, $C_E(S,\tau)$, in equation (3.23) is given by*

$$C_E(S,\tau)=\sum_{n=0}^{\infty}\frac{e^{-\lambda\tau}(\lambda\tau)^n}{n!}\mathbb{E}_{\mathbb{Q}}^{(n)}\{C_{BS}[SX_ne^{-\lambda k\tau},K,K,r,q,\tau,\sigma^2]\}, \tag{3.24}$$

where

$$\begin{aligned} C_{BS}[S,K,\beta,r,q,\tau,\sigma^2] &= Se^{-q\tau}\mathcal{N}[d_1(S,\beta,r,q,\tau,\sigma^2)] \\ &\quad-Ke^{-r\tau}\mathcal{N}[d_2(S,\beta,r,q,\tau,\sigma^2)], \\ d_1(S,\beta,r,q,\tau,\sigma^2) &= \frac{\ln\frac{S}{\beta}+\left(r-q+\frac{\sigma^2}{2}\right)\tau}{\sigma\sqrt{\tau}}, \\ d_2(S,\beta,r,q,\tau,\sigma^2) &= d_1(S,\beta,r,q,\tau,\sigma^2)-\sigma\sqrt{\tau}, \end{aligned} \tag{3.25}$$

$\mathcal{N}[\cdot]$ *is the cumulative normal density function, and we define* $X_n \equiv Y_1 Y_2 \ldots Y_n$ *and* $X_0 \equiv 1$, *along with*[6]

$$\mathbb{E}_{\mathbb{Q}}^{(n)}\{f(X_n)\} \equiv \int_0^\infty \int_0^\infty \ldots \int_0^\infty f(X_n) G(Y_1) G(Y_2) \ldots G(Y_n) dY_1 dY_2 \ldots dY_n$$
$$= \int_0^\infty f(X_n) G(X_n) dX_n.$$

The $Y_1, Y_2, \ldots, Y_n$ *are independent draws from the jump size distribution* $G(Y)$.

Proof. See Appendix 3.A. □

Equation (3.24) is of course Merton's 1976 solution for a European call option under jump-diffusion, with general jump size density $G(Y)$. Next we shall determine the early exercise premium $C_P(S, \tau)$.

Proposition 3.4. *The early exercise premium,* $C_P(S, \tau)$, *in equation* (3.23) *is given by*

$$C_P(S,\tau) = \sum_{n=0}^{\infty} \mathbb{E}_{\mathbb{Q}}^{(n)} \bigg\{ \int_0^\tau \frac{e^{-\lambda(\tau-\xi)}(\tau-\xi)^n}{n!}$$
$$\times [C_P^{(D)}[S X_n e^{-\lambda k(\tau-\xi)}, K, a(\xi), r, r, q, \tau-\xi, \sigma^2]$$
$$- \lambda C_P^{(J)}[S X_n e^{-\lambda k(\tau-\xi)}, K, a(\xi), r, q, \tau-\xi, \sigma^2; C^A(\cdot,\xi)]] d\xi \bigg\}, \tag{3.26}$$

where

$$C_P^{(D)}[S, K, a(\xi), R, r, q, \tau, \sigma^2]$$
$$= qSe^{-q\tau}\mathcal{N}[d_1(S, a(\xi), r, q, \tau, \sigma^2)]$$
$$- RKe^{-r\tau}\mathcal{N}[d_2(S, a(\xi), r, q, \tau, \sigma^2)], \tag{3.27}$$

$$C_P^{(J)}[S, K, a(\xi), r, q, \tau, \sigma^2; C^A(\cdot, \xi)]$$
$$= e^{-r\tau} \int_0^1 G(Y) \int_{a(\xi)}^{a(\xi)/Y} [C^A(\omega Y, \xi) - (\omega Y - K)]$$
$$\times \kappa(S, \omega, r, q, \tau, \sigma^2) d\omega dY, \tag{3.28}$$

[6]We assume that the density function G is of the form that facilitates the reduction of the n-dimensional integral to a single integral. This is true of the log-normal density function that we use later.

and

$$\kappa(S,\omega,r,q,\tau,\sigma^2) \equiv \frac{1}{\omega\sigma\sqrt{2\pi\tau}} \exp\left\{-\frac{1}{2}d_2^2(S,\omega,r,q,\tau,\sigma^2)\right\}. \tag{3.29}$$

The operator $\mathbb{E}_{\mathbb{Q}}^{(n)}$ *and functions* $\mathcal{N}$, d_1 *and* d_2 *have been defined in Proposition* 3.3.

Proof. See Appendix 3.B. □

On the left hand side in (3.28) we introduce the notation $C^A(\cdot,\xi)$ to indicate that the dependence on the option price is that of a functional (rather than of a function). It in fact has the form that

$$\int_0^1 \int_{a(\xi)}^{a(\xi)/Y} C^A(\omega Y,\xi)G(Y)g(\omega)d\omega dY, \tag{3.30}$$

for an appropriate function $g(\omega)$, a notation that will recur in many of the subsequent formulae.

Each of the linear terms in (3.26) represent discounted expected cash-flows incurred by the option holder when $S > a(\tau)$, as discussed previously for the interpretation of the inhomogeneous term in (3.9); the $C_P^{(D)}$ term is the expected value at time to maturity τ of the $qS - rK$ component in (3.9), and the $C_P^{(J)}$ term is the expected value of the integral term in (3.9). Combining C_E and C_P, we can now write down the integral equation for the American call option price, $C^A(S,\tau)$.

Proposition 3.5. *Substituting* (3.24) *and* (3.26) *into equation* (3.23), *the American call price,* $C^A(S,\tau)$, *is given by*

$$\begin{aligned} C^A(S,\tau) &= \sum_{n=0}^{\infty} \frac{e^{-\lambda\tau}(\lambda\tau)^n}{n!}\mathbb{E}_{\mathbb{Q}}^{(n)}\{C_{BS}[SX_n e^{-\lambda k\tau},K,K,r,q,\tau,\sigma^2]\},r \\ &\quad + \sum_{n=0}^{\infty} \mathbb{E}_{\mathbb{Q}}^{(n)}\left\{\int_0^\tau \frac{e^{-\lambda(\tau-\xi)}[\lambda(\tau-\xi)]^n}{n!}r\right. \\ &\quad \times [C_P^{(D)}[SX_n e^{-\lambda k(\tau-\xi)},K,a(\xi),r,r,q,\tau-\xi,\sigma^2] \\ &\quad \left. - \lambda C_P^{(J)}[SX_n e^{-\lambda k(\tau-\xi)},K,a(\xi),r,q,\tau-\xi,\sigma^2;C^A(\cdot,\xi)]]d\xi\right\}, \end{aligned} \tag{3.31}$$

where C_{BS} *is given by equation* (3.25), *and the functions* $C_P^{(D)}$, *and* $C_P^{(J)}$ *are given by equations* (3.27) *and* (3.28) *respectively.*

Proof. Direct substitution of (3.24) and (3.26) into (3.23) yields equation (3.31). □

Equation (3.31) is indeed an integral equation since the unknown option price also appears under the time integral on the right-hand side, in particular through the $C_P^{(J)}$ term. This is in contrast to the pure diffusion case where the equation corresponding to (3.31) (when $\lambda = 0$ and so that the $C_P^{(J)}$ term drops out) is simply an integral expression for the option price that can be evaluated once the free boundary has been determined.

The solution (3.31) is readily compared with that of Gukhal (2001), who derives (3.31) by generalising the compound option approach of Kim (1990) to the jump-diffusion case. The three additive components of the call value in equation (3.31) each have a clear economic interpretation, as outlined by Gukhal (2001). The first term, C_{BS}, represents the European component of the American call option's value, while the remaining two terms combine to form the total early exercise premium. The middle term is a natural extension of the early exercise premium that arises in the pure diffusion case. More specifically, this term calculates the dividend received when holding the underlying, less the interest payable on a loan of size K. Thus $C_P^{(D)}$ captures the potential income to the option holder should the option be exercised to buy the underlying by borrowing K at the risk-free rate. The third term, $C_P^{(J)}$, arises entirely due to the introduction of jumps in the price process for S, and captures the rebalancing costs incurred by the option holder whenever the price of the underlying jumps down from the stopping region into the continuation region as illustrated in Figure 3.1.

In equation (3.31), the value of the American call option is expressed as a function of the underlying asset price S, and time to maturity τ. As we have already noted, equation (3.31) also depends upon the unknown early exercise boundary, $a(\tau)$. By requiring the expression for $C^A(S, \tau)$ to satisfy the early exercise boundary condition (3.6), we can derive the integral equation for $a(\tau)$, namely

$$
\begin{aligned}
a(\tau) - K = & \sum_{n=0}^{\infty} \frac{e^{-\lambda\tau}(\lambda\tau)^n}{n!} \mathbb{E}_{\mathbb{Q}}^{(n)} \{ C_{BS}[a(\tau)X_n e^{-\lambda k \tau}, K, K, r, q, \tau, \sigma^2] \} \\
& + \sum_{n=0}^{\infty} \mathbb{E}_{\mathbb{Q}}^{(n)} \bigg\{ \int_0^{\tau} \frac{e^{-\lambda(\tau-\xi)}[\lambda(\tau-\xi)]^n}{n!} \\
& \times [C_P^{(D)}[a(\tau)X_n e^{-\lambda k(\tau-\xi)}, K, a(\xi), r, r, q, \tau-\xi, \sigma^2]
\end{aligned}
$$

$$- \lambda C_P^{(J)}[a(\tau)X_n e^{-\lambda k(\tau-\xi)}, K, a(\xi),$$
$$r, q, \tau - \xi, \sigma^2; C^A(\cdot, \xi)]]d\xi \Big\}. \tag{3.32}$$

Note that the integral equation (3.32) depends upon the unknown call value $C^A(S, \tau)$, which arises entirely from the integral terms that have been introduced by the presence of jumps in the dynamics for S.

The structure of the integral equation system consisting of (3.31) and (3.32) can be made more transparent by writing it succinctly as

$$C^A(S, \tau) = \Omega_C(S, \tau) + \int_0^\tau \Psi_C[a(\xi), \xi, \tau, S; C^A(\cdot, \xi)]d\xi, \tag{3.33}$$

$$a(\tau) = \Omega_a(a(\tau), \tau) + \int_0^\tau \Psi_a[a(\xi), \xi, \tau, a(\tau); C^A(\cdot, \xi)]d\xi, \tag{3.34}$$

where the definitions of the functions (Ω_C, Ψ_C) and (Ω_a, Ψ_a) can be inferred from the right hand sides of equations (3.31) and (3.32) respectively. The interdependence of (3.33) and (3.34) is obvious, and it is this interdependence that makes numerical implementation more involved than for the corresponding no-jump problem. There the dependence is sequential, that is first one solves for the free boundary, which then feeds into an integral expression for the option price. In fact equations (3.31)–(3.32) form a linked system of nonlinear Volterra integral equations of the second kind. Thus in order to implement these integral equations for the free boundary and call price, we need to develop numerical techniques to solve the linked integral equation system (3.31)–(3.32).

Before concluding this section, we present an alternative form for the double integral involving the function κ in equation (3.28) and defined in equation (3.29).

Proposition 3.6. *By changing the order of integration, C_P^J in equation (3.28) can be rewritten as*

$$C_P^{(J)}[S, K, a(\xi), r, q, \tau, \sigma^2; C^A(S, \xi)]$$
$$= e^{-r\tau} \int_0^1 [C^A(a(\xi)z, \xi) - (a(\xi)z - K)]$$
$$\times \int_0^z G(Y)\kappa(S/a(\xi), z, r, q, \tau, \sigma^2)dY dz. \tag{3.35}$$

Proof. The representation for $C_P^{(J)}$ in (3.28) cannot be further simplified without explicit knowledge of the density $G(Y)$. In cases where the density is known, however, it may be possible to complete the integration with respect to Y analytically. Here we change the order of integration to develop a form for the double integral that will be easier to evaluate using numerical integration methods. Recall from (3.28) that

$$C_P^{(J)}[S, K, a(\xi), r, q, \tau, \sigma^2; C^A(\cdot, \xi)]$$
$$= e^{-r\tau} \int_0^1 G(Y) \int_{a(\xi)}^{a(\xi)/Y} [C^A(\omega Y, \xi) - (\omega Y - K)]$$
$$\times \ \kappa(S, \omega, r, q, \tau, \sigma^2) d\omega dY.$$

Making the change of integration variable $z = \omega Y / a(\xi)$ we obtain

$$C_P^{(J)}[S, K, a(\xi), r, q, \tau, \sigma^2; C^A(\cdot, \xi)]$$
$$= e^{-r\tau} \int_0^1 \int_Y^1 G(Y)[C^A(a(\xi)z, \xi) - (a(\xi)z - K)]$$
$$\times \ \kappa(S/a(\xi), z, r, q, \tau, \sigma^2) dz dY.$$

Finally, changing the order of integration using Fubini's theorem, we obtain

$$C_P^{(J)}[S, K, a(\xi), r, q, \tau, \sigma^2; C^A(S, \xi)]$$
$$= e^{-r\tau} \int_0^1 [C^A(a(\xi)z, \xi) - (a(\xi)z - K)]$$
$$\times \int_0^z G(Y) \kappa(S/a(\xi), z, r, q, \tau, \sigma^2) dY dz.$$

□

From an economic point of view, the modified representation in (3.35) is less intuitive than the original, however it offers significant advantages when attempting to evaluate (3.28) numerically for specific forms of $G(Y)$. In particular, we will see in Sec. 3.5 that when $G(Y)$ is the log-normal density function given by Merton (1976), the innermost integral in (3.35) can be evaluated analytically. In this way we are able to reduce (3.35) to a one-dimensional integral, which makes the task of numerically evaluating (3.35) much simpler. We remind the reader that the $C_P^{(J)}$ term in turn must be integrated over time-to-maturity so that altogether the jump term would in this case involve the evaluation of a double integral.

3.4 Limit of the Early Exercise Boundary at Expiry

In order to implement numerical schemes we need to know the value of the free boundary just prior to expiry, at $\tau = 0^+$. Early literature, for instance

Amin (1993) and Carr and Hirsa (2003) simply assumed that this limit is identical to the corresponding pure diffusion case. However Chiarella and Ziogas (2009) show that this limit is in fact more subtle. They derive the limit by analysing the inhomogeneous term in the PIDE (3.9), and find that the presence of jumps does in fact have an impact on the early exercise boundary at expiry. This difference is expressed analytically in Proposition 3.7 below, and is based on the analysis of Wilmott *et al.* (1993). The simple, intuitive method used here is taken from Chiarella *et al.* (2004) who demonstrate that the approach of Wilmott *et al.* (1993) is equivalent to setting the inhomogeneous term in Jamshidian's 1992 form for the PDE to zero, setting $\tau = 0$, $S = a(0^+)$, and solving for the free boundary.

Proposition 3.7. *The limit of the early exercise boundary, $a(\tau)$, as $\tau \to 0^+$ is given by*

$$a(0^+) = K \max\left(1, \frac{r + \lambda \int_0^{K/a(0^+)} G(Y)dY}{q + \lambda \int_0^{K/a(0^+)} YG(Y)dY}\right). \tag{3.36}$$

Proof. Referring to the inhomogeneous PIDE (3.9), the inhomogeneous term of interest is[7]

$$H(S - a(\tau))\left\{qS - rK - \lambda \int_0^\infty [C^A(SY, \tau) - (SY - K)]G(Y)dY\right\}. \tag{3.37}$$

Setting the term in braces in (3.37) equal to zero and evaluating at $\tau = 0$ with $S = a(0^+)$ we have

$$qa(0^+) - rK - \lambda \int_0^\infty [C^A(a(0^+)Y, 0) - (a(0^+)Y - K)]G(Y)dY = 0. \tag{3.38}$$

Given that $C^A(S, 0) = (S - K)^+$, equation (3.38) becomes

$$qa(0^+) - rK - \lambda \int_0^\infty [(a(0^+)Y - K)^+ - (a(0^+)Y - K)]G(Y)dY = 0. \tag{3.39}$$

Since the integral term is zero for $Y \geq K/a(0^+)$, we have

$$qa(0^+) - rK + \lambda \int_0^{K/a(0^+)} (a(0^+)Y - K)G(Y)dY = 0, \tag{3.40}$$

[7]Note that since $C^A(SY, \tau) = SY - K$ for $Y \geq a(\tau)/S$, we have

$$\int_0^{a(\tau)/S} [C^A(SY, \tau) - (SY - K)]dY = \int_0^\infty [C^A(SY, \tau) - (SY - K)]dY.$$

which we can rearrange to give

$$a(0^+) = K\frac{r + \lambda\int_0^{K/a(0^+)} G(Y)dY}{q + \lambda\int_0^{K/a(0^+)} YG(Y)dY}. \tag{3.41}$$

Finally, by noting that $a(\tau) \geq K$ must hold[8] for all $\tau \geq 0$, we arrive at the result.

An alternative approach to the derivation of the limit of $a(\tau)$ is given by Kim (1990), in which he takes the limit of the integral equation for the free boundary as $\tau \to 0^+$. This approach is more involved than the one presented here, but we have verified that this approach leads to the same result. We refer the reader to Chiarella and Ziogas (2009) for further details. □

It is worthwhile to observe that when $\lambda = 0$ equation (3.36) simplifies to the limit derived by Kim (1990) for the pure diffusion American call free boundary. Note that (3.36) is an implicit expression for $a(0^+)$, but it can be solved quickly and accurately using standard root-finding techniques. Furthermore, as $q \to 0$ the solution to the implicit part of equation (3.36) increases without bound. Thus when $q = 0$ the $a(0^+)$ becomes infinite, and we observe the well-known property that it is never optimal to exercise an American call option early in the absence of dividends.

3.5 The American Call with Log-Normal Jumps

Before we begin exploring a numerical solution method for the integral equation system (3.31)–(3.32), we shall consider the specific example when the jump-size density, $G(Y)$, is a log-normal distribution as in the original model of Merton (1976). The probability density function for Y is given by

$$G(Y) = \frac{1}{Y\delta\sqrt{2\pi}}\exp\left\{-\frac{(\ln Y - (\gamma - \delta^2/2))^2}{2\delta^2}\right\}, \tag{3.42}$$

where we set $\gamma \equiv \ln(1 + k)$, and δ^2 is the variance of $\ln Y$. Furthermore we note that for this choice of $G(Y)$ we have $\mathbb{E}_{\mathbb{Q}}[Y] = e^{\gamma}$.

[8]This is true because it is never optimal to exercise a call option if $S < K$.

Proposition 3.8. *In the case where $G(Y)$ is given by equation (3.42), the integral equation for $C^A(S,\tau)$ in (3.31) becomes*

$$
\begin{aligned}
C^A(S,\tau) &= \sum_{n=0}^{\infty} \frac{e^{-\lambda'\tau}(\lambda'\tau)^n}{n!} C_{BS}[S,K,K,r_n(\tau),q,\tau,v_n^2(\tau)], \\
&+ \sum_{n=0}^{\infty} \left\{ \int_0^{\tau} \frac{e^{-\lambda'(\tau-\xi)}[\lambda'(\tau-\xi)]^n}{n!} \right. \\
&\times [C_P^{(D)}[S,K,a(\xi),r,r_n(\tau-\xi),q,\tau-\xi,v_n^2(\tau-\xi)] \\
&\left. - \lambda C_P^{(J)}[S,K,a(\xi),r_n(\tau-\xi),q,\tau-\xi,v_n^2(\tau-\xi);C^A(\cdot,\xi)]]d\xi \right\},
\end{aligned} \tag{3.43}
$$

where $\lambda' = \lambda(1+k)$, $r_n(\tau) = r - \lambda k + n\gamma/\tau$ and $v_n^2(\tau) = \sigma^2 + n\delta^2/\tau$.

Proof. The form equation (3.42) for $G(Y)$ implies that

$$
\begin{aligned}
\mathbb{E}_{\mathbb{Q}}^{(n)}\{f(X_n)\} = \int_0^{\infty} f(X_n) \frac{1}{X_n\delta\sqrt{2\pi n}} \\
\times \exp\left\{ -\frac{1}{2}\left(\frac{\ln X_n - n(\gamma - \frac{\delta^2}{2})}{\delta\sqrt{n}} \right)^2 \right\} dX_n.
\end{aligned} \tag{3.44}
$$

We shall use this to evaluate all of the $\mathbb{E}_{\mathbb{Q}}^{(n)}$ operators in equation (3.31).

1. **European Component**
   Using the results from Merton (1976), the European component becomes

$$
\begin{aligned}
&\sum_{n=0}^{\infty} \frac{e^{-\lambda\tau}(\lambda\tau)^n}{n!} \mathbb{E}_{\mathbb{Q}}^{(n)}\{C_{BS}[SX_n e^{-\lambda k\tau},K,K,r,q,\tau,\sigma^2]\} \\
&\quad = \sum_{n=0}^{\infty} \frac{e^{-\lambda'\tau}(\lambda'\tau)^n}{n!} C_{BS}[S,K,K,r_n(\tau),q,\tau,v_n^2(\tau)],
\end{aligned}
$$

   where $\lambda' = \lambda(1+k)$, $r_n(\tau) = r - \lambda k + n\gamma/\tau$, and $v_n^2(\tau) = \sigma^2 + n\delta^2/\tau$, with C_{BS} as defined in Proposition 3.3.

2. **Early Exercise Premium — First Term**
Consider the first part of the early exercise premium in equation (3.31), given by

$$C_P^{(1)}(S,\tau) = \sum_{n=0}^{\infty}\left\{\int_0^{\tau} \frac{e^{-\lambda(\tau-\xi)}[\lambda(\tau-\xi)]^n}{n!} \times \mathbb{E}_{\mathbb{Q}}^{(n)}\{C_P^{(D)}[SX_n e^{-\lambda k(\tau-\xi)}, K, a(\xi), r, r, q, \tau-\xi, \sigma^2]\}d\xi\right\}.$$

Using equation (3.27) for the definition of $C_P^{(D)}$ and equation (3.44), we can show that

$$\begin{aligned}&\mathbb{E}_{\mathbb{Q}}^{(n)}\{X_n\mathcal{N}[d_1(SX_n e^{-\lambda k(\tau-\xi)}, a(\xi), r, q, \tau-\xi, \sigma^2)]\}\\ &\quad = e^{n\gamma}\mathcal{N}[d_1(S, a(\xi), r_n(\tau-\xi), q, \tau-\xi, v_n^2(\tau-\xi))],\end{aligned}$$

and

$$\begin{aligned}&\mathbb{E}_{\mathbb{Q}}^{(n)}\{\mathcal{N}[d_2(SX_n e^{-\lambda k(\tau-\xi)}, a(\xi), r, q, \tau-\xi, \sigma^2)]\}\\ &\quad = \mathcal{N}[d_2(S, a(\xi), r_n(\tau-\xi), q, \tau-\xi, v_n^2(\tau-\xi))].\end{aligned}$$

Noting that $e^{n\gamma} = (k+1)^n$, $C_P^{(1)}$ becomes

$$C_P^{(1)}(S,\tau) = \sum_{n=0}^{\infty}\left\{\int_0^{\tau} \frac{e^{-\lambda'(\tau-\xi)}[\lambda'(\tau-\xi)]^n}{n!} \times C_P^{(D)}[S, K, a(\xi), r, r_n(\tau-\xi), q, \tau-\xi, v_n^2(\tau-\xi)]d\xi\right\}.$$

3. **Cost Term from Downward Jumps**
The final term to consider is the cost incurred when S jumps from the stopping region into the continuation region. From equation (3.42) this term is given by

$$C_P^{(2)}(S,\tau) = \lambda\sum_{n=0}^{\infty}\left\{\int_0^{\tau} \frac{e^{-\lambda(\tau-\xi)}[\lambda(\tau-\xi)]^n}{n!} \times \mathbb{E}_{\mathbb{Q}}^{(n)}\{C_P^{(J)}[SX_n e^{-\lambda k(\tau-\xi)}, K, a(\xi), r, q, \tau-\xi, \sigma^2; C^A(\cdot,\xi)]\}d\xi\right\}.$$

Referring to (3.28) and (3.29) for the definitions of $C_P^{(J)}$ and κ respectively, we find that in order to evaluate the the $\mathbb{E}_{\mathbb{Q}}^{(n)}$ operator, we must consider

$$\mathbb{E}_{\mathbb{Q}}^{(n)}\{\kappa(SX_n e^{-\lambda k(\tau-\xi)}, \omega, r, q, \tau-\xi, \sigma^2)\}$$
$$= \int_0^\infty \frac{1}{X_n \delta\sqrt{2\pi}} \exp\left\{-\frac{1}{2}\frac{[\ln X_n - n(\gamma - \frac{\delta^2}{2})]^2}{\delta^2 n}\right\} \frac{1}{\omega\sigma\sqrt{2\pi(\tau-\xi)}}$$
$$\times \exp\left\{\frac{-[(r-q-\lambda k - \frac{\sigma^2}{2})(\tau-\xi) + \ln\frac{SX_n}{\omega}]^2}{2\sigma^2(\tau-\xi)}\right\} dX_n.$$

Making the change of variable $x_n = \ln X_n$, this expectation can be evaluated as

$$\mathbb{E}_{\mathbb{Q}}^{(n)}\{\kappa(SX_n e^{-\lambda k(\tau-\xi)}, \omega, r, q, \tau-\xi, \sigma^2)\}$$
$$= \frac{1}{\omega\sqrt{2\pi(\tau-\xi)v_n^2(\tau-\xi)}}$$
$$\times \exp\left\{-\frac{[\ln\frac{S}{\omega} + (r_n(\tau-\xi) - q - \frac{v_n^2(\tau-\xi)}{2})(\tau-\xi)]^2}{2v_n^2(\tau-\xi)(\tau-\xi)}\right\}.$$

Finally, using the definitions for λ' and $r_n(\tau)$ we can rewrite $C_P^{(2)}$ as

$$C_P^{(2)}(S,\tau) = \lambda\sum_{n=0}^{\infty}\left\{\int_0^\tau \frac{e^{-\lambda'(\tau-\xi)}[\lambda'(\tau-\xi)]^n}{n!}\right.$$
$$\left.\times C_P^{(J)}[S, K, a(\xi), r_n(\tau-\xi), q, \tau-\xi, v_n^2(\tau-\xi); C^A(\cdot,\xi)]]d\xi\right\}.$$

Combining the results from points 1 to 3, we find that the integral equation for $C^A(S,\tau)$ in the case of log-normal jumps is given by equation (3.43) in Proposition 3.8. □

The last term of equation (3.43), which involves a double-integral, may be further simplified before implementing (3.43) numerically.

Proposition 3.9. *By use of Proposition* 3.6, *the term* $C_P^{(J)}$ *in Proposition* 3.8 *can be expressed as*

$$C_P^{(J)}[S, K, a(\xi), r_n(\tau), q, \tau, v_n^2(\tau); C^A(\cdot,\xi)]$$
$$= e^{-r_n(\tau)\tau}\int_0^1 [C^A(a(\xi)z,\xi) - (a(\xi)z - K)]$$
$$\times \kappa(S/a(\xi), z, r_{n+1}(\tau), q, \tau, v_{n+1}^2(\tau))$$
$$\times \mathcal{N}[D(S/a(\xi), z, r_n(\tau), q, v_n(\tau), v_{n+1}(\tau), \tau, \gamma, \delta)]dz, \qquad (3.45)$$

where

$$D(S/a(\xi), z, r_n(\tau), q, v_n(\tau), v_{n+1}(\tau), \tau, \gamma, \delta)$$
$$\equiv \frac{\delta^2 \ln\left(\frac{S}{a(\xi)z}\right) + \left[(\ln z)\, v_{n+1}^2(\tau) + \delta^2[r_n(\tau) - q] - \gamma v_n^2(\tau)\right]\tau}{v_n(\tau)v_{n+1}(\tau)\delta\tau}. \tag{3.46}$$

Proof. See Appendix 3.C. □

We note that in the form (3.45) the $C_P^{(J)}$ term in (3.43) now only involves a single integral, which results in a considerable saving in computational effort.

By use of (3.32), the integral equation for the early exercise boundary, $a(\tau)$, in the case of log-normal jump sizes, is given by

$$\begin{aligned} a(\tau) - K = &\sum_{n=0}^{\infty} \frac{e^{-\lambda'\tau}(\lambda'\tau)^n}{n!} C_{BS}[a(\tau), K, K, r_n(\tau), q, \tau, v_n^2(\tau)] \\ &+ \sum_{n=0}^{\infty}\left\{\int_0^{\tau} \frac{e^{-\lambda'(\tau-\xi)}[\lambda'(\tau-\xi)]^n}{n!}\right. \\ &\times [C_P^{(D)}[a(\tau), K, a(\xi), r, r_n(\tau-\xi), q, \tau-\xi, v_n^2(\tau-\xi)] \\ &- \lambda C_P^{(J)}[a(\tau), K, a(\xi), r_n(\tau-\xi), q, \tau-\xi, \\ &\left. v_n^2(\tau-\xi); C^A(\cdot,\xi)]]d\xi\right\}, \end{aligned} \tag{3.47}$$

where $C_P^{(D)}$ and $C_P^{(J)}$ are given by (3.27) and (3.45) respectively.

3.5.1 *Delta for the American call*

We now provide one further result regarding the delta of the American call option, $\Delta_C(S,\tau)$. This quantity is obviously important for hedging purposes, but it is also required by the numerical algorithm we consider in Sec. 3.7.

By differentiating (3.43) with respect to S, we find that

$$\Delta_C(S,\tau) = \sum_{n=0}^{\infty} \frac{e^{-\lambda'\tau}(\lambda'\tau)^n}{n!} \Delta_{BS}[S, K, K, r_n(\tau), q, \tau, v_n^2(\tau)]$$
$$+ \sum_{n=0}^{\infty}\left\{\int_0^{\tau} \frac{e^{-\lambda'(\tau-\xi)}[\lambda'(\tau-\xi)]^n}{n!}\right.$$

$$\times\,[\Delta_P^{(D)}[S,K,a(\xi),r,r_n(\tau-\xi),q,\tau-\xi,v_n^2(\tau-\xi)]$$
$$-\;\lambda\Delta_P^{(J)}[S,K,a(\xi),r_n(\tau-\xi),q,\tau-\xi,v_n^2(\tau-\xi);C^A(\cdot,\xi)]]d\xi\Big\},\tag{3.48}$$

where

$$\Delta_P^{(D)}[S,K,a(\xi),r,r_n(\tau),q,\tau,v_n^2(\tau)]$$
$$= e^{-q\tau}\Bigg\{\frac{1}{v_n(\tau)\sqrt{\tau}}\mathcal{N}'\left[d_1\left(S,a(\xi),r_n(\tau),q,\tau,v_n^2(\tau)\right)\right](q-r)$$
$$+\,q\mathcal{N}\left[d_1\left(S,a(\xi),r_n(\tau),q,\tau,v_n^2(\tau)\right)\right]\Bigg\},\tag{3.49}$$

$$\Delta_P^{(J)}[S,K,a(\xi),r_n(\tau),q,\tau,v_n^2(\tau);C^A(\cdot,\xi)]$$
$$= e^{-r_n(\tau)\tau}\int_0^1\frac{[C^A(a(\xi)z,\xi)-(a(\xi)z-K)]}{Sv_n(\tau)\sqrt{\tau}}$$
$$\times\,\kappa(S/a(\xi),z,r_{n+1}(\tau),q,\tau,v_{n+1}^2(\tau))$$
$$\times\left[\frac{\delta}{v_{n+1}(\tau)\sqrt{\tau}}\mathcal{N}'[D(S/a(\xi),z,r_n(\tau),q,v_n(\tau),v_{n+1}(\tau),\tau,\gamma,\delta)]\right.$$
$$\left.-\,d_2(S/a(\xi),z,r_{n+1}(\tau),q,\tau,v_{n+1}^2(\tau))\right]dz,\tag{3.50}$$

and we note that $\mathcal{N}'(x)=\exp(-x^2/2)/\sqrt{2\pi}$. Once we have found the price and free boundary for the American call option, we can readily evaluate (3.48) numerically to find delta.

3.6 Properties of the Free Boundary at Expiry

We consider the problem of analysing the behaviour of the early exercise boundary, $a(\tau)$, as $\tau\to 0^+$. To this end we can state the following proposition.

Proposition 3.10. *For $K,r>0$ and $q,\lambda\geq 0$ the equation*

$$x = K\frac{r+\lambda\int_0^{K/x}G(Y)dY}{q+\lambda\int_0^{K/x}YG(Y)dY}$$

has a solution x^* if and only if $q > 0$. This solution is unique. Moreover, $x^* > K$ if and only if

$$(r-q)+\int_0^1 (1-Y)G(Y)dY > 0.$$

Proof. Define

$$f(x) \equiv Kr + K\int_0^{K/x} G(Y)dY - x\left(q+\int_0^{K/x} YG(Y)dY\right),$$

then $f(x^*) = 0$. For $x \in (0,\infty)$ we find

$$f'(x) = -K^2/x^2G(K/x) - \left(q+\int_0^{K/x} YG(Y)dY\right) + K^2/x^2G(K/x)$$

$$= -\left(q+\int_0^{K/x} YG(Y)dY\right).$$

By inspection

$$\lim_{x\to 0} f(x) = K(r+\lambda).$$

Also,

$$\lim_{x\to\infty} f(x) = \lim_{x\to}(Kr-qx)$$

because

$$0 < x\int_0^{K/x} YG(Y)dY \le K\int_0^{K/x} G(Y)dY \to 0 \quad \text{as } x\to\infty.$$

If $q > 0$ then f is strictly decreasing from positive to negative values on $(0,\infty)$. Hence there exists a unique $x^* \in (0,\infty)$. We observe that

$$f(K) = K\left[(r-q)+\int_0^1 (1-Y)G(Y)dY\right].$$

If $f(K) \le 0$ then $x^* \in (0,K]$, otherwise $x^* \in (K,\infty)$. If $q = 0$ then $f(x) \to Kr$ as $x\to\infty$. Since $f' \le 0$ there is no x^* such that $f(x^*) = 0$.

Returning to (2.54) we see that for $q > 0$

$$a(0+) = K_{\max}(1, x^*).$$

We also note that

$$\partial f(x)/\partial q = -x < 0$$

so that f is increasing with decreasing q which implies that x^* increases as $q \to 0$ and that

$$\lim_{q\to 0} x^* = \infty.$$

Similarly, we find that $\partial f(x)/\partial\lambda > 0$ so that $a(0+)$ is increasing with λ. If $G(Y) > 0$ for $Y \in (0,\infty)$ so that

$$\int_0^{K/x} YG(Y)dY < \int_0^{K/x} K/xG(Y)dY.$$

□

Now we consider the special case when $G(Y)$ follows a log normal distribution.

Proposition 3.11. *When $G(Y)$ is given by the log-normal density* (3.42), *the limit of the early exercise boundary $a(\tau)$ as $\tau \to 0^+$ becomes*

$$a(0^+) = K \max\left(1, \frac{r + \lambda\mathcal{N}[\{\ln K/a(0^+) - (\gamma - \frac{\delta^2}{2})\}/\delta]}{q + \lambda'\mathcal{N}[\{\ln K/a(0^+) - (\gamma + \frac{\delta^2}{2})\}/\delta]}\right). \tag{3.51}$$

Proof. The proof follows by evaluating the integral terms in (3.36) using $G(Y)$ from (3.42). □

3.7 Numerical Implementation

We now provide a numerical scheme with which to evaluate the linked integral equation system for the option price and free boundary formed by (3.43) and (3.47). The proposed method is an extension of the quadrature scheme used by Kallast and Kivinukk (2003) for the pure diffusion case. Here we focus on the adjustments that are needed to deal with the introduction of jumps into the dynamics for S. We first discretise the time to maturity variable, τ, into N equally spaced intervals of length h. Thus $\tau = ih$ for $i = 0, 1, 2, \ldots, N$, and $h = T/N$. We denote the call price profile at time step i by $C^A(S, ih) = C_i(S)$, and similarly the free boundary at time step i by $a(ih) = a_i$. Using a standard numerical technique that is applied to Volterra integral equations, we can solve the system (3.43) and (3.47) for increasing values of i, until eventually the entire free boundary and price profile are found. When calculating the infinite summations, we continue adding terms until the size of the Poisson coefficient for a given value of n is less than 10^{-20}. For the parameter values considered here, this typically results in the use of around 30 terms in the summation. In order to start the algorithm we require the initial value of $C_0(S)$, which is simply the payoff function, and also a_0, where $a_0 \equiv a(0^+)$, which is given by equation (3.51).

Since the integral term in (3.45) depends upon $C^A(S, \tau)$, an approximation will be needed for $C_i(S)$. At each time step we have found that a suitable approximation is given by $C_{i-1}(S)$, which is simply the American call price at the previous time step. The price at the $(i-1)^{st}$ time step is calculated for a suitably large number of evenly-spaced S values. Here we use 25 points in the range $0 \leq S \leq 250$. All necessary interpolation is conducted using cubic splines fitted locally through 7 values of $C_{i-1}(S)$. We then use Newton's method to solve for the early exercise boundary, as in Kallast and Kivinukk (2003), with two necessary additions. The first addition addresses the evaluation of the inner integral (3.45) over the interval $[0,1]$. This is computed using Gaussian integration of moments, with parameter $\alpha = -0.5$. Full details of this Gauss quadrature scheme can be found in Abramowitz and Stegun (1970) (Chapter 25, p. 921). The second addition relates to finding the derivative of (3.47) with respect to $a(\tau)$ for use in Newton's method. This is given by (3.48) for $\Delta_C(S,\tau)$, evaluated at $S = a(\tau)$. Since it is difficult to determine the limit of the integrands in (3.48) as $\xi \to \tau$, we resolve this by taking the limits as $\xi \to \tau$ for the option price integrands in (3.43) and differentiating these with respect to $a(\tau)$, as is done by Kallast and Kivinukk (2003). These limits are all finite, including the new limit required for the jump-related integral term, and the required derivatives are easily determined. Since we need to evaluate the expression (3.48) for $\Delta_C(S,\tau)$ for use in Newton's method, there is no significant additional computation time required to evaluate the American call delta once the free boundary has been estimated.

Having determined the discretised forms for the price and delta of $C_i(S)$, we then use Newton's method to solve for a_i. Before proceeding to the next time step, we use a_i to calculate a new approximation for $C_i(S)$, which is required when evaluating the double integral term at all subsequent time steps. This update for C_i is essential to ensure that the estimated free boundary remains monotonic. Note that as the value of i increases, the computational burden will also increase at a "faster than linear" rate — the reason being that the integration at step i depends on all values of a_j and $C_j(S)$ for $j = 0, 1, 2, \ldots i-1$.

It should be noted that the proposed numerical scheme does not involve any iterations with respect to the approximation of the integral term. It is possible to improve the accuracy of the algorithm by updating the approximation for the integral term at the ith time step using the most recently computed estimates for $C_i(S)$ and a_i. In practice we have found that such an iteration does not add significantly to the accuracy of the results (up to the order of accuracy considered in Sec. 3.8), and that computation time is at least doubled by the introduction of the iteration process. Thus we have chosen not to iterate with respect to the integral term approximation.

To explore the efficiency of the proposed numerical integration method, we compare it with two alternative numerical methods. The first method involves a finite difference solution for the PIDE (3.3). We apply the Crank-Nicolson scheme to all terms except for the integral. We initially estimate the integral term by approximating $C_i(S)$ with the explicit approximation

$C_{i-1}(S)$, as in Carr and Hirsa (2003). We then evaluate the integral using the Hermite-Gauss quadrature scheme (see p. 890 and Table 25.10 on p. 924 of Abramowitz and Stegun (1970)). The resulting tridiagonal matrix is inverted using LU-decomposition, and the early exercise condition is then applied to the solution at each time step. An evenly spaced grid is used, and the free boundary is estimated at each time step using cubic spline interpolation of the price profile, combined with the bisection method.

To improve the accuracy of the Crank-Nicolson solution, we use a two-step procedure at each time step. After determining an initial solution at time step i, denoted here as $C_i^{(1)}(S)$, using the estimate of $C_{i-1}(S)$ in the integral term, we then find an updated estimate by repeating the solution process, now using the $C_i^{(1)}(S)$ values in the integral estimate. This provides a second approximation for the option price, which we denote by $C_i^{(2)}(S)$. In practice we find that $C_i^{(1)}$ typically converges from below, whilst $C_i^{(2)}$ converges from above. Thus we take $C_i(S) = (C_i^{(1)}(S) + C_i^{(2)}(S))/2$ for the final Crank-Nicolson solution. This appears to greatly improve the convergence rate for the Crank-Nicolson scheme, although we do not report details of the convergence of $C^{(1)}$ and $C^{(2)}$ here.[9] In all cases we set the S domain to be $0 \leq S \leq 250$. We also calculate the American call delta by taking a central difference approximation using the price estimates.

3.8 Numerical Results

To analyse the efficiency of the numerical integration method, we compute the price and delta of an American call option with 6-months to maturity, and a strike price of $K = 100$. The global variance[10] of returns, s^2, is set equal to 5.93%. The jump intensity is set to $\lambda = 1$, and the jump variance is $\delta^2 = 0.04$. We then consider six different parameter sets, specifically $\mathbb{E}_{\mathbb{Q}}[Y] = e^{\gamma}$ taking values of 0.95, 1.00 and 1.04, along with the combinations $r = 3\%, q = 5\%$ and $r = 5\%, q = 3\%$. Note that $\gamma > 0$ implies upward jumps on average, and $\gamma < 0$ implies downward jumps on average. When $\gamma = 0$, the expected price change from a jump is zero. The diffusion coefficient, σ^2,

[9]Briani *et al.* (2004) note that it is unclear how to select the stopping criteria when using iterative finite difference solutions for (3.3). Since we observe greater accuracy by using the average of the first and second iteration results than using the second iteration alone, the averaging scheme we use here seems more efficient than using a stopping criteria that involves three or more iterations.

[10]By the global variance we mean $\mathbb{E}[(dS/S - (\mu - \lambda k)dt)^2]$ calculated from equation (2.16) to be $\sigma^2 + \lambda[e^{2\gamma+\delta^2} - 2e^{\gamma} + 1]$ in the case of a log-normal jump density.

Table 3.1 Parameter values used for the diffusion variance and jump component. The global variance is fixed at $s^2 = 5.93\%$, determined by $s^2 = \sigma^2 + \lambda[e^{2\gamma+\delta^2} - 2e^{\gamma} + 1]$.

σ^2	λ	e^{γ}	δ^2
0.0593	0	—	—
0.0200	1.00	0.95	0.04
0.0185	1.00	1.00	0.04
0.0136	1.00	1.04	0.04

is chosen such that the global variance was preserved for varying values of γ. Table 3.1 summarises the values of σ^2 used to ensure that the global variance was the same for each combination of γ and λ.

We compute the root mean square error (RMSE) using option prices and deltas with $S = 80, 90, 100, 110$ and 120. This is repeated for each of the six parameter sets, from which the average runtime and RMSE are then calculated. For the integration method we use 20 integration points for the Gauss quadrature scheme, and consider a sequence of 10 different time step values, with $N = 10, 20, \ldots, 90, 100$.

For the Crank-Nicolson method the integral term is approximated using 50 integration points, and we again use 10 time step values, with $N = 50, 100, \ldots, 450, 500$. We set the number of space steps equal to double the number of time steps. The number of space steps is set to 5 times the number of time steps. The code for all methods has been implemented using LAHEY™FORTRAN 95 running on a PC with a Pentium 4 2.40 GHz processer, 512MB of RAM, and running the Windows XP Professional operating system.

In assessing the efficiency of the numerical integration method, we use a Crank-Nicolson solution with 10,000 time steps and 5,000 space steps for the true solution. Since the numerical integration scheme requires evaluation of the option delta as part of the solution, it is also of worthwhile to consider the efficiency with which delta is calculated. The true delta is estimated from the Crank-Nicolson solution using the central difference approximation for $(C_{i+1} - C_{i-1})/(S_{i+1} - S_{i-1})$. In both cases we find that for the values of S under consideration these methods are consistent to at least 4 decimal places, and thus we conclude that the Crank-Nicolson method using 10,000 time steps and 5,000 steps in the space variable provides a satisfactory estimate for the true price and delta.

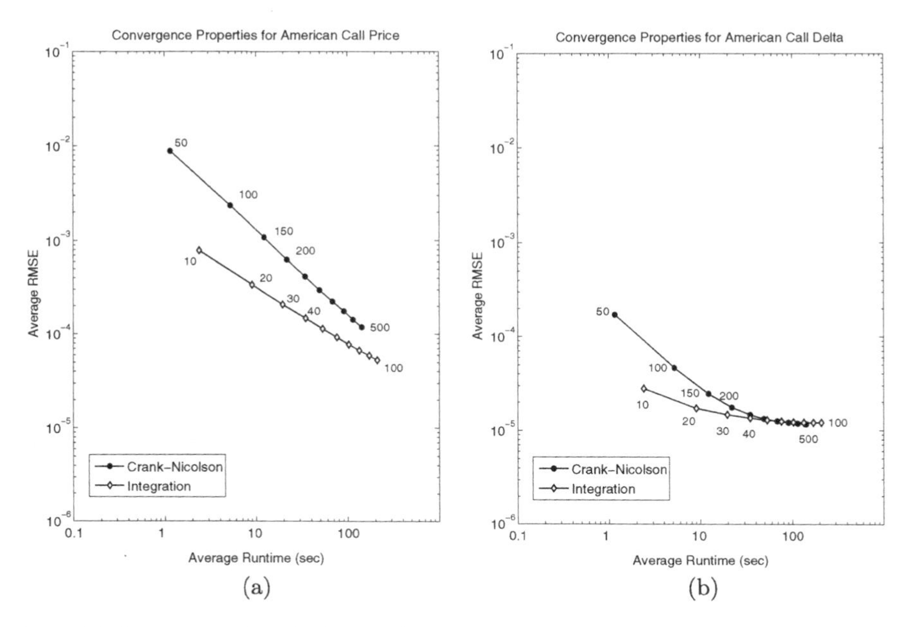

Fig. 3.2 Comparing the efficiency of numerical integration and the Crank-Nicolson scheme for the price and delta of American call options with log-normal jump sizes. We set $K = 100$, $T - t = 0.50$ and $\lambda = 1$. The RMSE is found using $S = 80, 90, 100, 110$ and 120. The average RMSE and runtimes have been determined using 6 parameter sets, with $r = 3\%$, $q = 5\%$, and $r = 5\%$, $q = 3\%$, along with $e^{\gamma} = 0.95, 1.00$ and 1.04. (a) Displays the price efficiency, whilst (b) shows the delta efficiency. Numbers on the plot indicate the time steps associated with a given point.

The relative efficiency of each method is shown in Figure 3.2 by comparing the RMSE as a function of runtime. Figure 3.2(a) shows the average RMSE error for the American call price, and Figure 3.2(b) displays the same information for the delta. Note that the average runtimes for each discretisation level are the same in Figures 3.2(a) and 3.2(b) since the price and delta were found using a single algorithm. First, we find that for the parameters and discretisations considered, the numerical integration method consistently displays greater efficiency than the Crank-Nicolson scheme for the American call price, and furthermore, numerical integration is more efficient than Crank-Nicolson when computing delta for computation times of up to 50 seconds.

We also note that the numerical integration algorithm has a slower rate of convergence for the delta relative to the other method. This could be attributed to cumulative numerical integration error as the number of time steps, and hence integration points, is increased. The rate of convergence for the numerical integration method is clearly better when computing the option price.

Thus from Figure 3.2[11] we conclude that of the two methods under consideration, we find that the numerical integration method consistently outperforms the Crank-Nicolson scheme when computing the option price, and is competitive with Crank-Nicolson when computing the delta for run-times of less than 50 seconds. Thus we can see that a simple extension of the numerical integration scheme presented by Kallast and Kivinukk (2003) to include log-normal jumps produces a numerical method that is comparable with the Crank-Nicolson scheme, however there is room for improvement in the computation of delta, in particular when the number of time steps is increased beyond 40. We also note that since the magnitude of the RMSE is much larger for the option price, the best way to select the optimal numerical method is to maximise the pricing efficiency whilst being aware of the RMSE for the deltas. For instance, for a runtime of five seconds, numerical integration is the most efficient.

Appendix

3.A Proof of Proposition 3.3

Consider the function $V_E(x,\tau)$, given by

$$V_E(x,\tau) = \mathcal{F}^{-1}\left\{\hat{V}(\eta,0)e^{-\left(\frac{1}{2}\sigma^2\eta^2+\phi i\eta+(r+\lambda)-\lambda A(\eta)\right)\tau}\right\}. \tag{3.52}$$

To evaluate this inversion, recall the convolution theorem for Fourier transforms given by

$$\mathcal{F}\left\{\int_{-\infty}^{\infty} f((x-u),\tau)g(u,\tau)du\right\} = \hat{F}(\eta,\tau)\hat{G}(\eta,\tau), \tag{3.53}$$

where $\hat{F}$ and $\hat{G}$ are the Fourier transforms, with respect to x, of $f(x,\tau)$ and $g(x,\tau)$ respectively. If we let $\hat{F}(\eta,\tau) = \exp(-(\frac{1}{2}\sigma^2\eta^2 + \phi i\eta + (r+\lambda) - \lambda A(\eta))\tau)$, then $f(x,\tau)$ is given by

$$f(x,\tau) = \frac{1}{2\pi}\int_{-\infty}^{\infty} e^{\lambda\tau A(\eta)}e^{-\left[\frac{1}{2}\sigma^2\eta^2\tau+i[\phi\tau+x]\eta+(r+\lambda)\tau\right]}d\eta.$$

Furthermore, let $\hat{G}(\eta,\tau) = \hat{V}(\eta,0)$, then $g(x,\tau)$ will simply be the payoff function, $g(x,0) = (e^x-1)^+$.

[11] Note that the reported runtimes indicate the total time required to find the free boundary, price and delta for the American call. Both axes are given in log-scale.

Using a Taylor series expansion for $e^{\lambda\tau A(\eta)}$, the expression for $f(x,\tau)$ becomes

$$f(x,\tau) = \frac{1}{2\pi}\int_{-\infty}^{\infty}\sum_{n=0}^{\infty}\frac{(\lambda\tau)^n A(\eta)^n}{n!}e^{-\left[\frac{1}{2}\sigma^2\eta^2\tau + i[\phi\tau + x]\eta + (r+\lambda)\tau\right]}d\eta. \quad (3.54)$$

Note that by definition

$$\begin{aligned} A(\eta)^n &= \left\{\int_0^{\infty} e^{-i\eta \ln Y} G(Y)dY\right\}^n \\ &= \int_0^{\infty} e^{-i\eta\ln Y_1}G(Y_1)dY_1 \ldots \int_0^{\infty} e^{-i\eta \ln Y_n}G(Y_n)dY_n \\ &= \int_0^{\infty}\int_0^{\infty}\cdots\int_0^{\infty} G(Y_1)G(Y_2)\ldots G(Y_n)e^{-i\eta\ln(Y_1Y_2\ldots Y_n)} \\ &\quad \times dY_1dY_2\ldots dY_n, \\ &= \mathbb{E}_{\mathbb{Q}}^{(n)}\left\{e^{-i\eta\ln X_n}\right\}, \end{aligned}$$

where

$$\mathbb{E}_{\mathbb{Q}}^{(n)}\{(\cdot)\} = \int_0^{\infty}\int_0^{\infty}\cdots\int_0^{\infty}(\cdot)G(Y_1)G(Y_2)\ldots G(Y_n)dY_1dY_2\ldots dY_n,$$

with $\mathbb{E}_{\mathbb{Q}}^{(0)}\{(\cdot)\} \equiv (\cdot)$, and $X_n \equiv Y_1Y_2\ldots Y_n$, with $X_0 \equiv 1$.

Substituting for $A(\eta)^n$ from equation (3.20), $f(x,\tau)$ becomes

$$f(x,\tau) = \sum_{n=0}^{\infty}\frac{(\lambda\tau)^n}{n!}\mathbb{E}_{\mathbb{Q}}^{(n)}\left\{\frac{1}{2\pi}\int_{-\infty}^{\infty} e^{-(r+\lambda)\tau}e^{-\left(\frac{1}{2}\sigma^2\eta^2\tau + i[\phi\tau + x + \ln X_n]\eta\right)}d\eta\right\}.$$

Recalling the result that

$$\int_{-\infty}^{\infty} e^{-p\xi^2 - q\xi}d\xi = \sqrt{\frac{\pi}{p}}e^{q^2/4p}, \quad (3.55)$$

we finally have the result that

$$\begin{aligned} \mathcal{F}^{-1}\{\hat{F}(\eta,\tau)\} &= f(x,\tau) \\ &= \sum_{n=0}^{\infty}\frac{e^{-\lambda\tau}(\lambda\tau)^n}{n!} \\ &\quad \times\mathbb{E}_{\mathbb{Q}}^{(n)}\left\{\frac{e^{-r\tau}}{\sigma\sqrt{2\pi\tau}}\exp\left\{-\frac{[x + \ln X_n + \phi\tau]^2}{2\sigma^2\tau}\right\}\right\}. \end{aligned} \quad (3.56)$$

Thus, by use of the convolution theorem (3.53) we have

$$V_E(x,\tau) = \sum_{n=0}^{\infty} \frac{e^{-\lambda\tau}(\lambda\tau)^n}{n!} \mathbb{E}_{\mathbb{Q}}^{(n)} \left\{ \frac{e^{-r\tau}}{\sigma\sqrt{2\pi\tau}} \int_0^{\infty} (e^u - 1) \right.$$

$$\left. \times \exp\left\{ -\frac{[x - u + \ln X_n + \phi\tau]^2}{2\sigma^2\tau} \right\} du \right\},$$

which, in terms of S is

$$C_E(S,\tau) = \sum_{n=0}^{\infty} \frac{e^{-\lambda\tau}(\lambda\tau)^n}{n!} \mathbb{E}_{\mathbb{Q}}^{(n)} \left\{ I_1(S,\tau) - I_2(S,\tau) \right\}, \quad (3.57)$$

where we set

$$I_1(S,\tau) \equiv \frac{e^{-r\tau}}{\sigma\sqrt{2\pi\tau}} \int_0^{\infty} K e^u \exp\left\{ -\frac{[\ln(SX_n/K) - u + \phi\tau]^2}{2\sigma^2\tau} \right\} du, \quad (3.58)$$

and

$$I_2(S,\tau) \equiv \frac{e^{-r\tau}}{\sigma\sqrt{2\pi\tau}} \int_0^{\infty} K \exp\left\{ -\frac{[\ln(SX_n/K) - u + \phi\tau]^2}{2\sigma^2\tau} \right\} du. \quad (3.59)$$

Beginning with I_1, we have

$$I_1(S,\tau) = \frac{Ke^{-r\tau}}{\sigma\sqrt{2\pi\tau}} e^{-\beta^2/(2\sigma^2\tau)} \int_0^{\infty} \exp\left\{ -\frac{u^2 - 2(\beta + \sigma^2\tau)u}{2\sigma^2\tau} \right\} du,$$

where $\beta \equiv \ln(SX_n/K) + \phi\tau$. Completing the square with respect to u and changing the integration variable, we find that (recall that $\phi = r - q - \lambda k - \sigma^2/2$)

$$I_1(S,\tau) = SX_n e^{-\lambda k\tau} e^{-q\tau} \mathcal{N}[d_1(SX_n e^{-\lambda k\tau}, K, r, q, \tau, \sigma^2)]. \quad (3.60)$$

For I_2, a suitable change of integration variable gives

$$I_2(S,\tau) = Ke^{-r\tau} \mathcal{N}[d_2(SX_n e^{-\lambda k\tau}, K, r, q, \tau, \sigma^2)]. \quad (3.61)$$

Finally, substituting I_1 and I_2 into (3.57), we find that

$$C_E(S,\tau) = \sum_{n=0}^{\infty} \frac{e^{-\lambda\tau}(\lambda\tau)^n}{n!} \mathbb{E}_{\mathbb{Q}}^{(n)} \left\{ C_{BS}[SX_n e^{-\lambda k\tau}, K, K, r, q, \tau, \sigma^2] \right\}, \quad (3.62)$$

where C_{BS} is the solution the Black-Scholes-Merton solution for a European call option.

3.B Proof of Proposition 3.4

Consider the function

$$V_P(x,\tau) = \int_0^\tau \mathcal{F}^{-1}\left\{\hat{F}_J(\eta,\xi)e^{-(\frac{\sigma^2\eta^2}{2}+i\phi\eta+(r+\lambda)-\lambda A(\eta))(\tau-\xi)}\right\}d\xi.$$

Using equation (3.19) we recall that

$$\mathcal{F}^{-1}\left\{\hat{F}_J(\eta,\xi)\right\} = F_J(x,\xi),$$

where F_J is defined by equation (3.55).

We use again the result (3.56) with τ replaced by $(\tau-\xi)$ to see that

$$\begin{aligned}
&\mathcal{F}^{-1}\left\{e^{-(\frac{\sigma^2\eta^2}{2}+i\phi\eta+(r+\lambda)-\lambda A(\eta))(\tau-\xi)}\right\} \\
&\quad = \sum_{n=0}^{\infty}\frac{e^{-\lambda(\tau-\xi)}[\lambda(\tau-\xi)]^n}{n!} \\
&\quad \times\mathbb{E}_{\mathbb{Q}}^{(n)}\left\{\frac{e^{-r(\tau-\xi)}}{\sigma\sqrt{2\pi(\tau-\xi)}}\exp\left\{-\frac{[x+\ln X_n+\phi(\tau-\xi)]^2}{2\sigma^2(\tau-\xi)}\right\}\right\}.
\end{aligned}$$

Thus by use of the convolution theorem (3.53) and equation (3.55) we obtain

$$\begin{aligned}
&V_P(x,\tau) \\
&\quad = \int_0^\tau\sum_{n=0}^{\infty}\left[\frac{e^{-\lambda(\tau-\xi)}[\lambda(\tau-\xi)]^n}{n!}\int_{-\infty}^{\infty}H(u-\ln b(\xi))\right. \\
&\quad \times\left(qe^u - r - \lambda\int_0^{b(\xi)e^{-x}}[V(u+\ln Y,\xi)-(Ye^u-1)]G(Y)dY\right) \\
&\quad \left.\times\mathbb{E}_{\mathbb{Q}}^{(n)}\left\{\frac{e^{-r(\tau-\xi)}}{\sigma\sqrt{2\pi(\tau-\xi)}}\exp\left\{-\frac{[x-u+\ln X_n+\phi(\tau-\xi)]^2}{2\sigma^2(\tau-\xi)}\right\}\right\}dud\xi\right],
\end{aligned}$$

which in terms of $S(=Ke^x)$ becomes (recall that $V_P(x,\tau)=C_P(S,\tau)/K$)

$$\begin{aligned}
C_P(S,\tau) &= \sum_{n=0}^{\infty}\frac{\lambda^n}{n!}\mathbb{E}_{\mathbb{Q}}^{(n)}\int_0^\tau(\tau-\xi)^n e^{-\lambda(\tau-\xi)} \\
&\quad \times[I_3(S,\tau)-I_4(S,\tau)-I_5(S,\tau)]d\xi, \qquad (3.63)
\end{aligned}$$

where

$$I_3(S,\tau) \equiv \frac{qe^{-r(\tau-\xi)}}{\sigma\sqrt{2\pi(\tau-\xi)}} \times \int_{\ln b(\xi)}^{\infty} Ke^u \exp\left\{-\frac{[\ln(SX_n/K) - u + \phi(\tau-\xi)]^2}{2\sigma^2(\tau-\xi)}\right\} du, \tag{3.64}$$

$$I_4(S,\tau) \equiv \frac{re^{-r(\tau-\xi)}}{\sigma\sqrt{2\pi(\tau-\xi)}} \times \int_{\ln b(\xi)}^{\infty} K \exp\left\{-\frac{[\ln(SX_n/K) - u + \phi(\tau-\xi)]^2}{2\sigma^2(\tau-\xi)}\right\} du, \tag{3.65}$$

and

$$I_5(S,\tau) \equiv \lambda\frac{e^{-r(\tau-\xi)}}{\sigma\sqrt{2\pi(\tau-\xi)}} \times \int_{\ln b(\xi)}^{\infty} \exp\left\{-\frac{[\ln(SX_n/K) - u + \phi(\tau-\xi)]^2}{2\sigma^2(\tau-\xi)}\right\} \times \int_0^{b(\xi)e^{-u}} [C^A(KYe^u,\xi) - (KYe^u - K)]G(Y)dY du. \tag{3.66}$$

To simplify I_3 and I_4, we make use of the results for I_1 and I_2 in the Proof of Proposition 3.3. Firstly, we note that (3.64) is simply (3.58) with τ replaced by $(\tau-\xi)$. Thus from (3.60) we have

$$I_3(S,\tau) = qSX_ne^{-\lambda k(\tau-\xi)}e^{-q(\tau-\xi)} \times \mathcal{N}[d_1(SX_ne^{-\lambda k(\tau-\xi)}, a(\xi), r, q, \tau-\xi, \sigma^2)]. \tag{3.67}$$

Similarly for I_2, we can use (3.61) to show that (3.65) is

$$I_4(S,\tau) = rKe^{-r(\tau-\xi)}\mathcal{N}[d_2(SX_ne^{-\lambda k(\tau-\xi)}, a(\xi), r, q, \tau-\xi, \sigma^2)]. \tag{3.68}$$

For I_5, we change the order of integration using Fubini's theorem, and make the change of integration variable $\omega = Ke^u$, which gives

$$I_5(S,\tau) = \lambda e^{-r(\tau-\xi)} \int_0^1 G(Y) \int_{a(\xi)}^{a(\xi)/Y} [C^A(\omega Y,\xi) - (\omega Y - K)] \times \kappa(SX_ne^{-\lambda k(\tau-\xi)}, \omega, r, q, \tau-\xi, \sigma^2)d\omega dY,$$

where κ is defined by (3.29). Finally, substituting I_3, I_4 and I_5 into (3.63) gives equation (3.26) from Proposition 3.4.

3.C Proof of Proposition 3.9

Referring to the result in Proposition 3.6, the integral term in equation (3.31) that we seek to evaluate is

$$I(S,z,\tau,\xi) \equiv \mathbb{E}_{\mathbb{Q}}^{(n)}\left\{\int_0^z G(Y)\kappa(SYX_n e^{-\lambda k(\tau-\xi)}/a(\xi), z, r, q, \tau-\xi, \sigma^2)\right\} dY$$

$$= \frac{1}{\delta\sqrt{2\pi}}\int_0^z \frac{1}{Y}\exp\left\{-\frac{1}{2}\left[\frac{\ln Y - (\gamma - \frac{\delta^2}{2})}{\delta}\right]^2\right\} J(S,z,\tau,\xi,Y)dY,$$

where

$$J(S,z,\tau,\xi,Y) \equiv \frac{1}{z\sigma\sqrt{2\pi(\tau-\xi)}}\frac{1}{\delta\sqrt{2\pi n}}$$

$$\times\int_0^\infty \frac{1}{X_n}\exp\left\{-\frac{1}{2}\left[\frac{\ln X_n - n(\gamma-\frac{\delta^2}{2})}{\delta\sqrt{n}}\right]^2\right\}$$

$$\times\exp\left\{-\frac{1}{2}\left[\frac{\ln\frac{SYX_n}{a(\xi)z} + (r-q-\lambda k - \frac{\sigma^2}{2})(\tau-\xi)}{\sigma\sqrt{\tau-\xi}}\right]^2\right\} dX_n.$$

To evaluate $I(S,z,\tau,\xi)$ we need to make use of the following integration result. Let α_1, α_2, β_1, β_2 and z be real-valued functions independent of the integration variable ω. Then by completing the square in the exponent, it can be shown that

$$\int_0^z \frac{1}{\omega}\exp\left\{-\frac{[\ln\omega+\beta_1]^2}{\alpha_1} - \frac{[\ln\omega+\beta_2]^2}{\alpha_2}\right\} d\omega$$

$$= \sqrt{\frac{\alpha_1\alpha_2\pi}{\alpha_1+\alpha_2}}\exp\left\{-\frac{(\beta_1-\beta_2)^2}{\alpha_1+\alpha_2}\right\}\mathcal{N}[f(z)], \tag{3.69}$$

where $f(z) = \sqrt{2}[(\alpha_1+\alpha_2)\ln z + \alpha_1\beta_2 + \alpha_2\beta_2]/\sqrt{\alpha_1\alpha_2(\alpha_1+\alpha_2)}$. Applying (3.69) to $J(S,z,\tau,\xi)$ we find that

$$J(S,z,\tau,\xi,Y) = \frac{1}{zv_n(\tau-\xi)\sqrt{2\pi(\tau-\xi)}}$$

$$\times\exp\left\{-\frac{[\ln\frac{SY}{a(\xi)z} + (r_n(\tau-\xi) - q - \frac{v_n^2(\tau-\xi)}{2})(\tau-\xi)]^2}{2v_n^2(\tau-\xi)(\tau-\xi)}\right\},$$

where $r_n(\tau)$ and $v_n(\tau)$ are given by Proposition 3.8, and thus $I(S, z, \tau, \xi)$ becomes

$$I(S, z, \tau, \xi) = \frac{1}{z v_n(\tau - \xi)\sqrt{2\pi(\tau - \xi)}}$$
$$\times \frac{1}{\delta\sqrt{2\pi}} \int_0^z \frac{1}{Y} \exp\left\{ -\frac{[\ln Y - (\gamma - \frac{\delta^2}{2})]^2}{2\delta^2} \right\}$$
$$\times \exp\left\{ -\frac{[\ln \frac{SY}{a(\xi)z} + (r_n(\tau - \xi) - q - \frac{v_n^2(\tau-\xi)}{2})(\tau - \xi)]^2}{2v_n^2(\tau - \xi)(\tau - \xi)} \right\} dY.$$

Finally, we again apply (3.69) to $I(S, z, \tau, \xi)$ and obtain

$$I(S, z, \tau, \xi) = \frac{1}{z v_{n+1}(\tau - \xi)\sqrt{2\pi(\tau - \xi)}} \mathcal{N}[f(z)]$$
$$\times \exp\left\{ -\frac{[\ln \frac{S}{a(\xi)z} + (r_{n+1}(\tau - \xi) - q - \frac{v_{n+1}^2(\tau-\xi)}{2})(\tau - \xi)]^2}{2v_{n+1}^2(\tau - \xi)(\tau - \xi)} \right\},$$

where

$$f(z) = \frac{\delta^2 \ln \frac{S}{za(\xi)} + \left[(\ln z) v_{n+1}^2(\tau - \xi) + \delta^2 [r_n(\tau - \xi) - q] - \gamma v_n^2(\tau - \xi)\right](\tau - \xi)}{v_n(\tau - \xi) v_{n+1}(\tau - \xi) \delta (\tau - \xi)}$$
$$\equiv D(S/a(\xi), z, r_n(\tau - \xi), q, v_n(\tau - \xi), v_{n+1}(\tau - \xi), \tau - \xi, \gamma, \delta).$$

Substituting for $I(S, z, \tau, \xi)$ into (3.31) and combining this with the results in Proposition 3.8, we arrive at equation (3.45) of Proposition 3.9.

Chapter 4

American Option Prices under Stochastic Volatility and Jump-Diffusion Dynamics — The Transform Approach

4.1 Introduction

In this chapter we consider the problem of pricing American options under the combined stochastic volatility and jump-diffusion model of Bates (1996). We focus here on the representation of the solution from two perspectives, integral transforms and the probability approach. The integral transform approach involves identifying the homogeneous partial-integro differential equation (PIDE) for the American call price in a restricted domain, and using the technique by Jamshidian (1992) to convert this to an inhomogeneous PIDE in an unrestricted domain. We then apply the Fourier transform technique, in conjunction with results from Cheang *et al.* (2013), to derive a linked system of integral equations for the price and early exercise surface of the American call. With regard to the probability approach, this was developed by Karatzas (1988) when the underlying follows pure diffusion dynamics and extended by Pham (1997) to the case of jump-diffusion dynamics. Here we shall extend this approach to the situation when the underlying follows a stochastic volatility and jump-diffusion dynamics.

The remainder of this chapter is structured as follows. Sec. 4.2 outlines the free boundary problem that arises from pricing an American call option under stochastic volatility and jump-diffusion dynamics. Sec. 4.3 applies Jamshidian's method to derive an inhomogeneous PIDE for the American call price, which is then solved using a Fourier transform approach. In this way we obtain the linked integral equation system for the early exercise surface and American call price. Sec. 4.4 approaches the problem by using the martingale representation of the American option as in Karatzas (1988) and

Pham (1997). Concluding remarks are presented in Sec. 4.5. Most lengthy mathematical derivations are given in appendices.

4.2 The Problem Statement — The Merton-Heston Model

Let $C^A(S, v, \tau)$ be the price of an American call option written on a stock of price S with time to expiry τ and strike price K. For the underlying dynamics, we assume that the stochastic differential equation (SDE) for S is given by the jump-diffusion process proposed by Merton (1976), in conjunction with the square root volatility process by Heston (1993). Thus the dynamics for S are governed by the SDE system

$$dS = (\mu - \lambda k)Sdt + \sqrt{v}SdZ_1 + (Y-1)SdN, \tag{4.1}$$

$$dv = \kappa_v(\theta - v)dt + \sigma_v\sqrt{v}dZ_2. \tag{4.2}$$

In (2.1), μ is the instantaneous return per unit time, v is the instantaneous squared volatility per unit time, and Z_1 is a standard Wiener process. Furthermore, we define

$$dN = \begin{cases} 1, & \text{with probability} \quad \lambda dt, \\ 0, & \text{with probability} \quad (1 - \lambda dt), \end{cases}$$

and set

$$k = \mathbb{E}_{\mathbb{Q}}[(Y-1)] = \int_0^\infty (Y-1)G(Y)dY,$$

where $G(Y)$ is the continuous probability density function for the multiplicative jump sizes, Y, generated by the measure $\mathbb{Q}$. In (4.2), θ is the long-run mean for v, κ_v is the rate of mean reversion, σ_v is the instantaneous volatility of v per unit time, and Z_2 is a standard Wiener process correlated with Z_1 such that $\mathbb{E}[dZ_1 dZ_2] = \rho dt$. Note that dN, Y, dZ_1 and dZ_2 are otherwise uncorrelated.

We have seen in Chapter 2 that the problem of pricing an option under these conditions reduces to solving the partial-integro differential equation (PIDE) (2.12) subject to the boundary conditions (2.13), (2.14) and (2.15) and the initial condition (2.19). We note that the PIDE (2.12) may also be written in the form (2.16).

The free boundary value problem defined by (2.12)–(2.15) involves a homogeneous PIDE in a restricted S-domain. There are several approaches that can be used to solve this PIDE, and here we extend the method used by Jamshidian (1992) for the standard American option pricing problem. Thus rather than solving (2.12)–(2.15) in the restricted domain $0 \leq S \leq a(v, \tau)$, we derive an equivalent inhomogeneous PIDE which is to be solved in the unrestricted domain $0 \leq S < \infty$. The inhomogeneous term is determined by evaluating (2.12) in the stopping region.

Proposition 4.1. *The free boundary value problem (2.12)–(2.15) for $C^A(S, v, \tau)$ in the restricted domain $0 < S \leq a(v, \tau)$ is equivalent to solving the inhomogeneous PIDE*

$$\begin{aligned} \frac{\partial C^A}{\partial \tau} &= \frac{vS^2}{2}\frac{\partial^2 C^A}{\partial S^2} + \rho\sigma_v vS\frac{\partial^2 C^A}{\partial S \partial v} + \frac{\sigma_v^2 v}{2}\frac{\partial^2 C^A}{\partial v^2} \\ &\quad + (r - q - \lambda^* k^*)S\frac{\partial C^A}{\partial S} + (\kappa_v[\theta - v] - \lambda_v v)\frac{\partial C^A}{\partial v} - rC^A \\ &\quad + \lambda^* \int_0^\infty [C^A(SY, v, \tau) - C^A(S, v, \tau)]G^*(Y)dY + f(S, v, \tau), \end{aligned} \tag{4.3}$$

in the unrestricted domain $0 \leq S < \infty$, $0 \leq v < \infty$, $0 \leq \tau \leq T$, subject to the initial condition (2.19) and boundary conditions (2.13)–(2.15), and where we define the jump-risk adjusted Poisson intensity, λ^*, and jump size distribution $G^*(Y)$, under the jump measure $\mathbb{Q}^*$ according to

$$\begin{aligned} \lambda^* \mathbb{E}_{\mathbb{Q}^*}[F(Y)] &= \lambda^* \int_0^\infty F(Y)G^*(Y)dY \\ &\equiv \lambda \int_0^\infty F(Y)(1 - \lambda_J(Y))G(Y)dY, \end{aligned} \tag{4.4}$$

for a sufficiently well-behaved function F and where

$$k^* \equiv \mathbb{E}_{\mathbb{Q}^*}[(Y - 1)]. \tag{4.5}$$

The inhomogeneous term, $f(S, v, \tau)$, is given by

$$\begin{aligned} f(S, v, \tau) &= \mathbb{1}_{\{S - a(v,\tau) \geq 0\}} \\ &\quad \times \left\{ qS - rK - \lambda^* \int_0^{a(v,\tau)/S} \right. \\ &\quad \left. \times [C^A(SY, v, \tau) - (SY - K)]G^*(Y)dY \right\}, \end{aligned} \tag{4.6}$$

where $\mathbb{1}_{\{x \geq 0\}}$ is the indicator function defined by

$$\mathbb{1}_{\{x \geq 0\}} = \begin{cases} 1, & x \geq 0, \\ 0, & x < 0. \end{cases} \tag{4.7}$$

Proof. Refer to Appendix 4.A. □

There are several advantages in solving (4.3) rather than (2.12). The first is that the solution to the inhomogeneous PIDE can be found directly by first solving the corresponding homogeneous PIDE, and then applying Duhamel's principle to complete the solution for a given inhomogeneous term $f(S, v, \tau)$. The second advantage is that solving the homogeneous form of (4.3) can be readily achieved using several techniques. Here we will use a Fourier transform approach, as this will allow us to readily derive the solution for the stochastic volatility and jump-diffusion case by applying results for the American call under stochastic volatility given by Adolfsson, Chiarella, Ziogas and Ziveyi (2013). The other major benefit of this approach is that the solution is naturally decomposed into the sum of the corresponding European option price plus the early exercise premium. This follows because the solution to the homogeneous PIDE is in fact the value of the European call written on S.

In Figure 4.1 we give a schematised version of the relationship between Jamshidian's approach and McKean's approach showing how they are related to each other.

We now provide the details for Duhamel's principle, as it applies to the inhomogeneous PIDE (4.3).

Proposition 4.2. *According to Duhamel's principle, the solution to the inhomogeneous PIDE* (4.3) *is given by*

$$C^A(S, v, \tau) = C_E(S, v, \tau) + C_P(S, v, \tau), \tag{4.8}$$

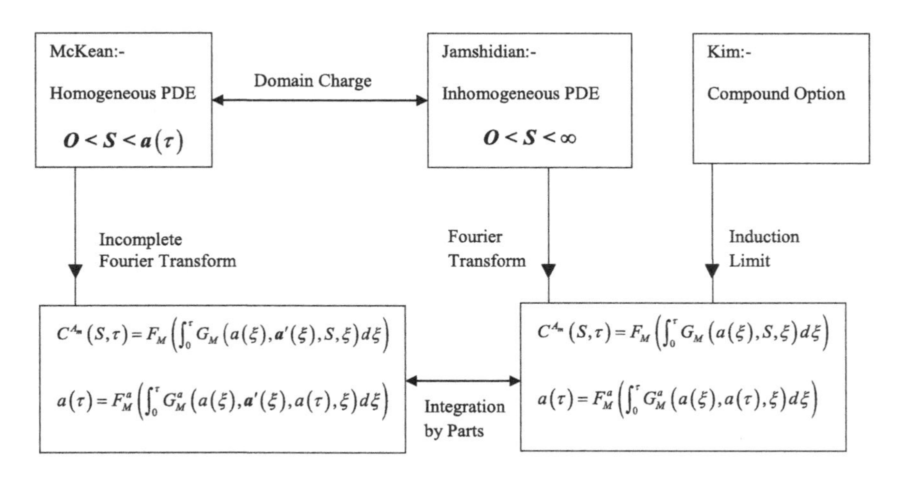

Fig. 4.1 The different approaches to American option pricing.

where

$$C_E(S,v,\tau) \equiv e^{-r\tau}\int_0^\infty \int_{-\infty}^\infty c(e^y,w)\bar{G}(y,w,\tau;S,v)dydw, \tag{4.9}$$

and

$$C_P(S,v,\tau) \equiv \int_0^\tau e^{-r(\tau-\xi)}\int_0^\infty \int_{-\infty}^\infty f(e^y,w,\xi)\bar{G}(y,w,\tau-\xi;S,v)dydwd\xi. \tag{4.10}$$

The initial condition $C^A(S,v,0) = c(S,v)$ is given by (2.19), and the inhomogeneous term $f(S,v,\tau)$ is given by (4.6). The function $\bar{G}(y,w,\tau;S,v)$ is the joint transition probability density function[1] *for S and v.*

Proof. Refer to Appendix 4.B. □

Note that (4.9) is readily identifiable as the price of the corresponding European call option written on S, and thus (4.10) is the early exercise premium for $C^A(S,v,\tau)$. Hence by solving the homogeneous version of (4.3) for a general payoff function $c(S,v)$, we will be able to infer the function $\bar{G}(y,w,\tau;S,v)$,[2] and this is all we require to find the early exercise premium for the American call option.

4.3 The Integral Transform Solution

In applying Proposition 4.2 to solve the PIDE (4.3) subject to (2.12)–(2.15), we need to solve the homogeneous form of (4.3) for the European call option price, $C_E(S,v,\tau)$, which satisfies the homogeneous PIDE

$$\begin{aligned}\frac{\partial C_E}{\partial \tau} &= \frac{vS^2}{2}\frac{\partial^2 C_E}{\partial S^2} + \rho\sigma_v vS\frac{\partial^2 C_E}{\partial S\partial v} + \frac{\sigma_v^2 v}{2}\frac{\partial^2 C_E}{\partial v^2}\\ &\quad + (r-q-\lambda^* k^*)S\frac{\partial C_E}{\partial S} + (\kappa_v[\theta - v] - \lambda_v v)\frac{\partial C_E}{\partial v} - rC_E\\ &\quad + \lambda^*\int_0^\infty [C_E(SY,v,\tau) - C_E(S,v,\tau)]G^*(Y)dY, \end{aligned}\tag{4.11}$$

subject to the initial condition (2.19). Here we shall use the Fourier transform to reduce the two-dimensional PIDE to a one-dimensional PDE whose

---

[1]The function $\bar{G}(y,u,\tau;S,v)$ can also be interpreted as the Green's function associated with the PIDE (4.3). For more on the theory of Green's functions, we refer the reader to the extensive literature on PDEs, such as Duffy (2001).

[2]We can also find the PIDE for $\bar{G}(y,w,\tau;S,v)$ and solve this subject to the initial condition $\bar{G}(y,w,0;S,v) = \delta(\ln S - y)\delta(v-w)$. Here we will focus on the more general problem of solving the homogeneous PIDE (4.3) for a general initial condition, since the steps involved are essentially the same.

solution is already known, thereby allowing us to readily find the solution to (4.11).

We begin by making a change of variable. Let $S = e^x$ and $C_E(S, v, \tau) = e^{-r\tau}U(x, v, \tau)$, so that equation (4.11) becomes

$$\begin{aligned}\frac{\partial U}{\partial \tau} &= \frac{v}{2}\frac{\partial^2 U}{\partial x^2} + \rho\sigma_v v\frac{\partial^2 U}{\partial x \partial v} + \frac{\sigma_v^2 v}{2}\frac{\partial^2 U}{\partial v^2} \\ &\quad + \left(r - q - \lambda^* k^* - \frac{v}{2}\right)\frac{\partial U}{\partial x} + (\alpha - \beta v)\frac{\partial U}{\partial v} \\ &\quad + \lambda^* \int_0^\infty [U(x + \ln Y, v, \tau) - U(x, v, \tau)]G^*(Y)dY, \end{aligned} \tag{4.12}$$

which is to be solved in the region $-\infty < x < \infty$, $0 \leq v < \infty$, $0 \leq \tau \leq T$, subject to the initial condition $U(x, v, 0) = (e^x - K)^+$, with $\alpha \equiv \kappa_v\theta$ and $\beta \equiv \kappa_v + \lambda_v$. Since x does not appear in the coefficients for any of the terms in (4.12), we are now able to take the Fourier transform of the PIDE with respect to x, which we present in Proposition 4.3.

It should be noted that integral transforms require knowledge of the behaviour of the functions being transformed at the extremities of the domain. In Chapter 3 we discussed how using a technique introduced by Carr and Madan (1999) we were able to overcome this problem. Of course in Chapter 3 we were only considering jump-diffusion processes but not stochastic volatility. Here we are considering stochastic volatility. The same remarks apply as applied in Chapter 3 and we shall not discuss it further here.

Proposition 4.3. *Let $\mathcal{F}_x\{U(x, v, \tau)\}$ be the Fourier transform of $U(x, v, \tau)$ taken with respect to x. Define the Fourier transform as*

$$\mathcal{F}_x\{U(x, v, \tau)\} = \int_{-\infty}^{\infty} e^{i\phi x}U(x, v, \tau)dx = \hat{U}(\phi, v, \tau). \tag{4.13}$$

The Fourier transform of (4.12) is the solution of

$$\frac{\partial \hat{U}}{\partial \tau} = \frac{\sigma_v^2 v}{2}\frac{\partial^2 \hat{U}}{\partial v^2} + (\alpha - \Theta v)\frac{\partial \hat{U}}{\partial v} + \left(\frac{\Lambda}{2}v - i\phi\Psi\right)\hat{U}, \tag{4.14}$$

where

$$\begin{aligned}\Theta &= \Theta(\phi) \equiv \beta + \rho\sigma_v i\phi, \\ \Lambda &= \Lambda(\phi) \equiv i\phi - \phi^2, \\ \Psi &= \Psi(\phi) \equiv r - q - \lambda^* k^* - \frac{\lambda^*}{i\phi}(A(\phi) - 1),\end{aligned}$$

and

$$A(\phi) \equiv \int_0^\infty G^*(Y) e^{-i\phi \ln Y} dY. \tag{4.15}$$

We note that the initial condition $\hat{U}(\phi, v, 0) \equiv \hat{u}(\phi, v)$ can be obtained by calculating directly $\mathcal{F}_x\{U(x, v, 0)\}$.

Proof. Refer to Appendix 4.C. □

The general solution of the two-dimensional PIDE in the case of no-jumps (when the PIDE becomes a simple PDE) has already been derived by Feller (1951) using a Laplace transform approach in the v direction. The solution procedure in the case of stochastic volatility only is given in Adolfsson *et al.* (2013).

In applying the Laplace transform in the v direction we must make certain assumptions about the behaviour of $\hat{U}(\phi, v, \tau)$. Specifically, we assume that $e^{-sv}\hat{U}(\phi, v, \tau)$ and $e^{-sv}\partial\hat{U}(\phi, v, \tau)/\partial v$ tend to zero as $v \to \infty$, where s is the Laplace transform variable.[3] Proposition 4.4 provides the Laplace transform of (4.13) with respect to v.

Proposition 4.4. *Let $\mathcal{L}_v\{\hat{U}(\phi, v, \tau)\}$ be the Laplace transform of $\hat{U}(\phi, v, \tau)$ taken with respect to v, defined as*

$$\mathcal{L}_v\{\hat{U}(\phi, v, \tau)\} = \int_0^\infty e^{-sv}\hat{U}(\phi, v, \tau) dv \equiv \tilde{U}(\phi, s, \tau). \tag{4.16}$$

Applying the Laplace transform we find that $\tilde{U}$ satisfies the PDE

$$\frac{\partial \tilde{U}}{\partial \tau} + \left(\frac{\sigma_v^2 s^2}{2} - \Theta s + \frac{\Lambda}{2}\right)\frac{\partial \tilde{U}}{\partial s} = \left[(\alpha - \sigma_v^2)s + \Theta - i\phi\Psi\right]\tilde{U} + f_L(\tau), \tag{4.17}$$

with initial condition $\tilde{U}(\phi, s, 0) \equiv \tilde{u}(\phi, v)$, and where $f_L(\tau) \equiv (\sigma_v^2 s^2/2 - \alpha)\hat{U}(\phi, 0, \tau)$ is determined such that

$$\lim_{s \to \infty} \tilde{U}(\phi, s, \tau) = 0. \tag{4.18}$$

Proof. Refer to Appendix 4.D. □

---

[3]Again we can adopt the technique of Carr and Madan (1999) to justify these assumptions or adopt the approach used in the physical sciences, which is to assume that the solution is as given and then to verify that the solution obtained works.

Equation (4.17) is now a first order PDE which can be solved using the method of characteristics. The unknown function $f_L(\tau)$ on the right hand side of (4.17) is then found by applying the condition (4.18). In this way we are able to solve (4.17) for $\tilde{U}(\phi, s, \tau)$, and the result is given now in Proposition 4.5.

Proposition 4.5. *Using the method of characteristics, and subsequently applying condition* (4.18) *to determine* $f_L(\tau)$, *the solution to the first order PDE* (4.17) *is*

$$\begin{aligned}\tilde{U}(\phi, s, \tau) = \exp&\left\{\left[\frac{(\alpha - \sigma_v^2)(\Theta - \Omega)}{\sigma_v^2} + \Theta - i\phi\Psi\right]\tau\right\}\\ &\times\left(\frac{2\Omega}{(\sigma_v^2 s - \Theta + \Omega)(e^{\Omega\tau} - 1) + 2\Omega}\right)^{2-\frac{2\alpha}{\sigma_v^2}} \int_0^\infty \hat{u}(\phi, w)e^{-\left(\frac{\Theta-\Omega}{\sigma_v^2}\right)w}\\ &\times\exp\left\{\frac{-2\Omega w(\sigma_v^2 s - \Theta + \Omega)e^{\Omega\tau}}{\sigma_v^2[(\sigma_v^2 s - \Theta + \Omega)(e^{\Omega\tau} - 1) + 2\Omega]}\right\}\\ &\times\Gamma\left(\frac{2\alpha}{\sigma_v^2} - 1; \frac{2\Omega w e^{\Omega\tau}}{\sigma_v^2(e^{\Omega\tau} - 1)}\right.\\ &\left.\times\frac{2\Omega}{(\sigma_v^2 s - \Theta + \Omega)(e^{\Omega\tau} - 1) + 2\Omega}\right) dw,\end{aligned} \tag{4.19}$$

where

$$\Omega = \Omega(\phi) \equiv \sqrt{\Theta^2(\phi) - \sigma_v^2\Lambda(\phi)}, \tag{4.20}$$

$\Gamma(n; z)$ *is a (lower) incomplete gamma function, defined as*

$$\Gamma(n; z) = \frac{1}{\Gamma(n)}\int_0^z e^{-\xi}\xi^{n-1}d\xi, \tag{4.21}$$

and $\Gamma(n)$ *is the (complete) gamma function given by*

$$\Gamma(n) = \int_0^\infty e^{-\xi}\xi^{n-1}d\xi. \tag{4.22}$$

Proof. Refer to Appendix 4.E. □

Having determined $\tilde{U}(\phi, s, \tau)$, we now seek to return to the original variables S and v, and thus obtain the general solution for $C^A(S, v, \tau)$. We begin this process by inverting the Laplace transform in Proposition 4.6, again using the techniques of Feller (1951).

Proposition 4.6. *The inverse Laplace transform of* $\tilde{U}(\phi, s, \tau)$ *in* (4.19) *is*

$$\hat{U}(\phi, v, \tau) = \int_0^\infty \hat{u}(\phi, w) e^{\frac{(\Theta-\Omega)}{\sigma_v^2}(v-w+\alpha\tau)} e^{-i\phi(\Psi)\tau}$$

$$\times \frac{2\Omega e^{\Omega\tau}}{\sigma_v^2(e^{\Omega\tau}-1)} \left(\frac{we^{\Omega\tau}}{v}\right)^{\frac{\alpha}{\sigma_v^2}-\frac{1}{2}} \exp\left\{\frac{-2\Omega}{\sigma_v^2(e^{\Omega\tau}-1)}(we^{\Omega\tau}+v)\right\}$$

$$\times I_{\frac{2\alpha}{\sigma_v^2}-1}\left(\frac{4\Omega}{\sigma_v^2(e^{\Omega\tau}-1)}(wve^{\Omega\tau})^{\frac{1}{2}}\right) dw, \tag{4.23}$$

where $I_k(z)$ *is the modified Bessel function of the first kind, defined as*

$$I_k(z) = \sum_{n=0}^{\infty} \frac{\left(\frac{z}{2}\right)^{2n+k}}{\Gamma(n+k+1)n!}. \tag{4.24}$$

Proof. Refer to Appendix 4.F. □

Next we take the inverse Fourier transform of equation (4.23) and return to the original variables, allowing us to identify the form of the Green's function $\bar{G}(y, w, \tau; S, v)$ associated with $C^A(S, v, \tau)$.

Proposition 4.7. *Given the definition of* $\mathcal{F}_x$ *in* (4.13), *the inverse Fourier transform is*

$$\mathcal{F}_x^{-1}\{\hat{U}(\phi, v, \tau)\} = \frac{1}{2\pi}\int_{-\infty}^{\infty} e^{-i\phi x}\hat{U}(\phi, v, \tau)d\phi = U(x, v, \tau). \tag{4.25}$$

The inverse Fourier transform of (4.23), *in terms of* S *and* $C_E(S, v, \tau)$, *turns out to be*

$$C_E(S, v, \tau) = e^{-r\tau}\int_0^\infty \int_{-\infty}^{\infty} (e^y - K)^+ \bar{G}(y, w, \tau; S, v)dydw, \tag{4.26}$$

where $\bar{G}(y, w, \tau; S, v)$ *is given by*

$$\bar{G}(y, w, \tau; S, v) = \sum_{n=0}^{\infty} \frac{\lambda^* e^{-\lambda^*\tau}}{n!} \mathbb{E}_{\mathbb{Q}^*}^{(n)}$$

$$\times \left\{\frac{1}{2\pi}\int_{-\infty}^{\infty} e^{i\phi y} e^{\frac{(\Theta-\Omega)}{\sigma_v^2}(w-v+\alpha\tau)} e^{-i\phi(r-q-\lambda^* k^*)\tau}\right.$$

$$\times e^{-i\phi \ln SX_n} \frac{2\Omega}{\sigma_v^2(e^{\Omega\tau}-1)} \left(\frac{we^{\Omega\tau}}{v}\right)^{\frac{\alpha}{\sigma_v^2}-\frac{1}{2}}$$

$$\times \exp\left\{\frac{-2\Omega}{\sigma_v^2(e^{\Omega\tau}-1)}(we^{\Omega\tau}+v)\right\}$$

$$\left.\times I_{\frac{2\alpha}{\sigma_v^2}-1}\left(\frac{4\Omega}{\sigma_v^2(e^{\Omega\tau}-1)}(wve^{\Omega\tau})^{\frac{1}{2}}\right) d\phi\right\}, \tag{4.27}$$

with $X_n \equiv Y_1 Y_2 \ldots Y_n$, $X_0 \equiv 1$, *and*

$$\begin{aligned}\mathbb{E}_{\mathbb{Q}^*}^{(n)}[f(X_n)] &= \int_0^\infty f(X_n) G^*(X_n) dX_n \\ &= \int_0^\infty \int_0^\infty \cdots \int_0^\infty f(Y_1 Y_2 \ldots Y_n) \\ &\quad \times G^*(Y_1) G^*(Y_2) \ldots G^*(Y_n) dY_1 dY_2 \ldots dY_n. \end{aligned} \tag{4.28}$$

Note that each Y_j, $j = 0, \ldots, n$ *is an independent jump drawn from the density* $G^*(Y)$, *and it is assumed that the density* $G^*(Y)$ *is of a form which facilitates the reduction from an* n*-dimensional integral to a one-dimensional integral in* (4.28).[4]

Proof. Refer to Appendix 4.G. □

Having found the function $\bar{G}(y, w, \tau; S, v)$ required by Duhamel's principle in Proposition 4.2 we can now find both the European call price and the early exercise premium term for the American call option. Firstly, we simplify the European call price (4.26), and express it in a form that is analogous to the solution of Heston (1993).

Proposition 4.8. *Carrying out the integration with respect to* w *in* (4.26), *the European call price,* $C_E(S, v, \tau)$, *becomes*

$$\begin{aligned} C_E(S, v, \tau) &= \sum_{n=0}^{\infty} \frac{(\lambda^* \tau)^n e^{-\lambda^* \tau}}{n!} \\ &\quad \times \mathbb{E}_{\mathbb{Q}^*}^{(n)} \{ S X_n e^{-\lambda^* k^* \tau} e^{-q\tau} P_1^H (S X_n e^{-\lambda^* k^* \tau}, v, \tau; K) \\ &\quad - K e^{-r\tau} P_2^H (S X_n e^{-\lambda^* k^* \tau}, v, \tau; K) \}, \end{aligned} \tag{4.29}$$

where

$$P_j^H(S, v, \tau; K) = \frac{1}{2} + \frac{1}{\pi} \int_0^\infty Re \left(\frac{f_j(S, v, \tau; \phi) e^{-i\phi \ln K}}{i\phi} \right) d\phi, \tag{4.30}$$

for $j = 1, 2$, *with*

$$f_j(S, v, \tau; \phi) = \exp\{ B_j(\phi, \tau) + D_j(\phi, \tau) v + i\phi \ln S \}, \tag{4.31}$$

$$B_j(\phi, \tau) = i\phi(r - q)\tau + \frac{\alpha}{\sigma_v^2} \left\{ (\Theta_j + \Omega_j)\tau - 2 \ln \left(\frac{1 - Q_j e^{\Theta_j \tau}}{1 - Q_j} \right) \right\},$$

$$D_j(\phi, j) = \frac{(\Theta_j + \Omega_j)}{\sigma_v^2} \left(\frac{1 - e^{\Omega_j \tau}}{1 - Q_j e^{\Omega_j \tau}} \right),$$

[4] This holds true for certain popular types of distributions, such as lognormal (Merton (1976)) and double exponential (Kou (2002)).

and $Q_j = (\Theta_j + \Omega_j)/(\Theta_j - \Omega_j)$, *where we define* $\Theta_1 = \Theta(i - \phi)$, $\Omega_1 = \Omega(i - \phi)$, *and* $\Theta_2 = \Theta(-\phi)$, $\Omega_2 = \Omega(-\phi)$.

Proof. Refer to Appendix 4.H. □

Next we use (4.10) to determine the early exercise premium for the American call. The integral term in (4.6) gives rise to an additional component in the early exercise premium that measures rebalancing costs incurred by the option holder when the underlying asset price, S, jumps from the stopping region, $S > a(v, \tau)$, back into the continuation region.

Proposition 4.9. *By use of equations* (4.6) *and* (4.27), *the early exercise premium for the American call,* $C_P(S, v, \tau)$ *in* (4.10) *can be expressed as*

$$C_P(S, v, \tau) = C_P^D(S, v, \tau) - \lambda^* C_P^J(S, v, \tau). \tag{4.32}$$

The term $C_P^D(S, v, \tau)$ *is defined as*

$$\begin{aligned} &C_P^D(S, v, \tau)\\ &\equiv \sum_{n=0}^{\infty} \frac{(\lambda^*\tau)^n e^{-\lambda^*\tau}}{n!} \mathbb{E}_{\mathbb{Q}^*}^{(n)} \left\{ \int_0^{\tau} \int_0^{\infty} (\tau - \xi)^n e^{-\lambda^*(\tau-\xi)} \right.\\ &\quad \times [qSX_n e^{-\lambda^*(\tau-\xi)} e^{-q(\tau-\xi)} P_1^A(SX_n e^{-\lambda^*(\tau-\xi)}, v, \tau - \xi; w, a(w, \xi))\\ &\quad \left. - rKe^{-r(\tau-\xi)} P_2^A(SX_n e^{-\lambda^*(\tau-\xi)}, v, \tau - \xi; w, a(w, \xi))] dw d\xi \right\}, \end{aligned} \tag{4.33}$$

where

$$\begin{aligned} &P_j^A(S, v, \tau - \xi; w, a(w, \xi))\\ &= \frac{1}{2} + \frac{1}{\pi} \int_0^{\infty} Re\left(\frac{g_j(S, v, \tau - \xi; \phi, w) e^{-i\phi \ln a(w,\xi)}}{i\phi} \right) d\phi, \end{aligned} \tag{4.34}$$

for $j = 1, 2$, *with*

$$\begin{aligned} g_j(S, v, \tau - \xi; \phi, w) &= e^{\frac{(\Theta_j - \Omega_j)}{\sigma_v^2}(v - w + \alpha(\tau-\xi))} e^{i\phi(r-q)(\tau-\xi)} e^{i\phi \ln S}\\ &\quad \times \frac{2\Omega_j}{\sigma_v^2(e^{\Omega j(\tau-\xi)} - 1)} \left(\frac{w e^{\Omega_j(\tau-\xi)}}{v} \right)^{\frac{\alpha}{\sigma_v^2} - \frac{1}{2}}\\ &\quad \times \exp\left\{ \frac{-2\Omega_j}{\sigma_v^2(e^{\Omega j(\tau-\xi)} - 1)} (w e^{\Omega_j(\tau-\xi)} + v) \right\}\\ &\quad \times I_{\frac{2\alpha}{\sigma_v^2} - 1}\left(\frac{4\Omega_j}{\sigma_v^2(e^{\Omega_j(\tau-\xi)} - 1)} (wv e^{\Omega j(\tau-\xi)})^{\frac{1}{2}} \right), \end{aligned} \tag{4.35}$$

$I_k(z)$ *given by* (4.24), *and* Θ_j *and* Ω_j *are given in Proposition* 4.8.

The term $C_P^J(S,v,\tau)$ is defined as

$$C_P^J(S,v,\tau) = \sum_{n=0}^{\infty} \frac{(\lambda^*\tau)^n e^{-\lambda^*\tau}}{n!}$$
$$\times \mathbb{E}_{\mathbb{Q}^*}^{(n)} \left\{ \int_0^\tau \int_0^\infty (\tau-\xi)^n e^{-\lambda^*(\tau-\xi)} e^{-r(\tau-\xi)} \right.$$
$$\times \int_0^1 G(Y) \int_{\ln a(w,\xi)}^{\ln[a(w,\xi)/Y]} [C^A(zY,w,\xi) - (zY-K)]$$
$$\left. \times \bar{Q}_J(z,w,\tau-\xi; SX_n e^{-\lambda^* k^*(\tau-\xi)}, v) dz dY dw d\xi \right\}, \quad (4.36)$$

where

$$\bar{Q}_J(z,w,\tau-\xi; SX_n e^{-\lambda^* k^*(\tau-\xi)}, v)$$
$$\equiv \frac{1}{2\pi z} \int_{-\infty}^{\infty} e^{\frac{(\Theta-\Omega)}{\sigma_v^2}(v-w+\alpha(\tau-\xi))} e^{-i\phi(r-q)(\tau-\xi)}$$
$$\times e^{-i\phi \ln(SX_n e^{-\lambda^* k^*(\tau-\xi)}/z)} \frac{2\Omega}{\sigma_v^2(e^{\Omega(\tau-\xi)}-1)} \left(\frac{w e^{\Omega(\tau-\xi)}}{v}\right)^{\frac{\alpha}{\sigma_v^2}-\frac{1}{2}}$$
$$\times \exp\left\{ \frac{-2\Omega}{\sigma_v^2(e^{\Omega\tau}-1)} (w e^{\Omega(\tau-\xi)} + v) \right\}$$
$$\times I_{\frac{2\alpha}{\sigma_v^2}-1} \left(\frac{4\Omega}{\sigma_v^2(e^{\Omega(\tau-\xi)}-1)} (wv e^{\Omega(\tau-\xi)})^{\frac{1}{2}} \right) d\phi. \quad (4.37)$$

Proof. Refer to Appendix 4.I. □

Each linear component of the early exercise premium in (4.32) has a distinct financial interpretation. The first part, C_P^D, denotes the component of the early exercise premium arising from the diffusion part of the dynamics for S. Specifically, C_P^D is the expected present value of the portfolio $qS - rK$ held whenever S is in the stopping region. The second term, $\lambda^* C_P^J$, arises from the presence of jumps,[5] and is the expected present value of the cost incurred by the option holder whenever S jumps from the stopping region back into the continuation region. A detailed explanation of this rebalancing cost is given by Chiarella and Ziogas (2006).

[5]Note that when there are no jumps in the model, $\lambda^* = 0$ and C_P^J no longer contributes towards the early exercise premium.

We also note that the functions $f_j(S, v, \tau; \phi)$ in (4.31) and $g_j(S, v, \tau; \phi, w)$ in (4.35) are related according to

$$f_j(S, v, \tau; \phi) = \int_0^\infty g_j(S, v, \tau; \phi, w) dw.$$

Having derived integral representations for the European call price and early exercise premium, we now obtain the integral equation for the American call price. We can also readily derive the corresponding integral equation for the early exercise boundary, and thus determine the linked system of integral equations for $C^A(S, v, \tau)$ and $a(v, \tau)$.

Proposition 4.10. *The price of an American call, $C^A(S, v, \tau)$, written on S is given by*

$$C^A(S, v, \tau) = C_E(S, v, \tau) + C_P^D(S, v, \tau) - \lambda^* C_P^J(S, v, \tau), \tag{4.38}$$

where $C_E(S, v, \tau)$ is given by (4.29), and the terms $C_P^D(S, v, \tau)$ and $C_P^J(S, v, \tau)$ are given respectively by (4.33) and (4.36). Equation (4.38) depends upon the early exercise boundary, $a(v, \tau)$, which is the solution to the integral equation

$$C^A(a(v, \tau), v, \tau) = a(v, \tau) - K. \tag{4.39}$$

Proof. Substituting (4.29) and (4.32) into (4.8) gives (4.38). Evaluating (4.38) at $S = a(v, \tau)$ and applying boundary condition (2.14) produces (4.39). □

Equations (4.38)–(4.39) both contain integrals involving $C^A(S, v, \tau)$ and $a(v, \tau)$. The dependence upon C^A arises because of the presence of the jump terms. This means that one cannot solve sequentially for $a(v, \tau)$ and $C^A(S, v, \tau)$, as in the corresponding situation when jumps are not present. While it is possible to develop numerical methods that reduce this dependence, as demonstrated by Chiarella and Ziogas (2006) for the case of American options under jump-diffusion dynamics, such approaches involve an exponentially increasing computational burden as the number of underlying stochastic factors in the model increases.

Tzavalis and Wang (2003) provide an integration method for pricing American call options under stochastic volatility. One of the features of this method is that the free boundary is approximated as an exponential-linear function of v, which in turn provides a reduction of the dimensions for the integration in $C_P^D(S, v, \tau)$. This will not hold true once jumps are introduced, as the term $C_P^J(S, v, \tau)$ cannot be simplified in this manner. We note, however, that depending on the functional form of $G^*(Y)$, we may be able to complete the integration with respect to Y analytically in $C_P^J(S, v, \tau)$, after interchanging the order of integration for z and Y. This, combined with the need to evaluate the infinite sums arising from the Poisson arrival process for the jumps, results in the fact that the system (4.38)–(4.39) would be very cumbersome to solve.

4.4 The Martingale Representation

In this section we derive the representation of the option value by using the probabilistic arguments originally applied to the American option pricing problem by Karatzas (1988) and extended to the stochastic volatility case by Touzi (1999) and the jump-diffusion case by Pham (1997).

Assume a filtered probability measure space $(\Omega, \mathcal{F}, \{\mathcal{F}_t\}, \mathbb{P})$, where $\mathbb{P}$ is the market measure, and the filtration $\{\mathcal{F}_t\}$ generates all the relevant processes required in our model. Let $C^E(S, v, t)$ (note that we write t rather than τ to emphasize that the dependence on t is important) be the price of a European call option at time t written on a stock of price S_t and strike price K. The price of its American counterpart is denoted by $C^A(S, v, t)$. For the underlying dynamics, we assume that the stochastic differential equation (SDE) for S is given by the jump-diffusion process proposed by Merton (1976), in conjunction with the square root volatility process suggested by Heston (1993). Thus the dynamics for S_t are governed by the SDE system (2.1) and (2.2).

The return jump-size has moment generating function in the $\mathbb{P}$ given by

$$M_{\mathbb{P},J}(u) = \mathbb{E}_{\mathbb{P}}[e^{uJ}].$$

It will also be convenient to introduce the correlation matrix

$$\Sigma_v = \begin{pmatrix} 1 & \rho \\ \rho & 1 \end{pmatrix}.$$

We introduce the vector notation $\boldsymbol{Z}_t = (Z_{1,t}, Z_{2,t})^\top$ for the Wiener processes, and $\boldsymbol{\zeta}_t = (\zeta_{1,t}, \zeta_{2,t})^\top$ for the market prices of Z_1 and Z_2 risk. Note that $\zeta_{2,t}$ is referred to as the market price of volatility risk.

In order to facilitate the analysis, a Radon-Nikodým derivative is needed for the transformation of measures from the original market measure $\mathbb{P}$ to some equivalent measure $\mathbb{Q}$.

Proposition 4.11. *Let $\mathbb{P}$ and $\mathbb{Q}$ be equivalent probability measures. Consider the probability measure space $(\Omega, \mathcal{F}, \{\mathcal{F}_t\}, \mathbb{P})$ such that $\{\mathcal{F}_t\}$ is the* *natural filtration generated by correlated Wiener components* (Z_1, Z_2) *and a compound Poisson process* $\sum_{n=0}^{N_t} J_n$ *and a Radon-Nikodým derivative*

$$\left.\frac{d\mathbb{Q}}{d\mathbb{P}}\right|_t = e^{-\int_0^t (\Sigma_v^{-1}\boldsymbol{\zeta}_u)^\top d\boldsymbol{Z}_u - \frac{1}{2}\int_0^t \boldsymbol{\zeta}_u^\top \Sigma_v^{-1}\boldsymbol{\zeta}_u du}\, e^{\sum_{i=1}^{N_t}(\gamma J_i + v) - \lambda\kappa' t}, \tag{4.40}$$

where

$$\kappa' = e^v M_{\mathbb{P},J}(\gamma) - 1,$$

$\gamma \in \mathbb{R}$, $\nu \in \mathbb{R}$ *and* $\boldsymbol{\zeta}_u \in \mathbb{R}^2$. *Then the Wiener components* Z_i *have drift* $\zeta_{i,t}$ *under the measure* $\mathbb{Q}$ *and the compound process process* $\sum_{n=0}^{N_t} J_n$ *has a new intensity rate* $\widetilde{\lambda} = \lambda(1 + \kappa')$ *and a new distribution for the jump sizes given by the moment generating function*

$$M_{\mathbb{Q},J}(u) = \frac{M_{\mathbb{P},J}(u+\gamma)}{M_{\mathbb{P},J}(\gamma)}.$$

Proof. See Runggaldier (2003). □

The equivalent martingale measure $\mathbb{Q}$ is chosen so that $\left\{\frac{S_t e^{qt}}{e^{rt}}\right\}$ is a $\mathbb{Q}$-martingale once v, γ and $\zeta_{2,t}$ in the Radon-Nikodým derivative equation (4.40) in Proposition 4.11 are chosen.

For convenience, define $C_{t-}^E = C_t^E(S_{t-}, v, t)$ as the pre-jump option value. Standard application of Itô's Lemma for jump-diffusion processes yields the option price dynamics

$$\begin{aligned} dC_t^E = &\left[\frac{\partial C_{t-}^E}{\partial t} + (\mu - \lambda\kappa)S_{t-}\frac{\partial C_{t-}^E}{\partial S} + \kappa_v(\theta - v_t)\frac{\partial C_{t-}^E}{\partial v} + \frac{v_t S_{t-}^2}{2}\frac{\partial^2 C_{t-}^E}{\partial S^2}\right.\\ &\left. + \rho\sigma_v v_t S_{t-}\frac{\partial^2 C_{t-}^E}{\partial S\partial v} + \frac{\sigma_v^2 v_t}{2}\frac{\partial^2 C_{t-}^E}{\partial v^2}\right]dt + \sqrt{v_t}S_{t-}\frac{\partial C_{t-}^E}{\partial S}dZ_{1,t}\\ &+ \sigma_v\sqrt{v_t}\frac{\partial C_{t-}^E}{\partial v}dZ_{2,t} + [C_t^E(S_{t-}e^J, v, t) - C_{t-}^E(S_{t-}, v, t)]dN_t. \end{aligned} \tag{4.41}$$

If $\zeta_{2,t}$ is specified, then by Girsanov's theorem for Wiener process, there exists

$$d\widetilde{Z}_{2,t} = \zeta_{2,t}dt + dZ_{2,t}$$

such that $\widetilde{Z}_{2,t}$ is standard Brownian motion under the $\mathbb{Q}$-measure.

Therefore, the dynamics for the volatility process equation (4.2) becomes

$$\begin{aligned} dv_t &= \kappa_v(\theta - v_t)dt - \zeta_{2,t}\sqrt{v_t}\sigma_v dt + \sigma_v\sqrt{v_t}d\widetilde{Z}_{2,t}\\ &= [\kappa_v(\theta - v_t) - \lambda_v v_t]dt + \sigma_v\sqrt{v_t}d\widetilde{Z}_{2,t}, \end{aligned} \tag{4.42}$$

if we choose

$$\zeta_{2,t} = \frac{\lambda_v \sqrt{v_t}}{\sigma_v}$$

to coincide with Heston's choice of the market price of volatility risk.

From Proposition 4.11, the Poisson arrival process N_t has intensity $\widetilde{\lambda} = \lambda(1+\kappa')$ and the jump-size J has moment generating function given by

$$M_{\mathbb{Q},J}(u) = \frac{M_{\mathbb{P},J}(u+\gamma)}{M_{\mathbb{P},J}(\gamma)},$$

so that expected relative jump-size increment under the $\mathbb{Q}$-measure is

$$\widetilde{\kappa} = \mathbb{E}_{\mathbb{Q}}[e^J] - 1.$$

Now the dynamics of $S_t e^{qt}$, measured in units of the money market account, are given by

$$d\left(\frac{S_t e^{qt}}{e^{rt}}\right) = \left(\frac{S_{t-} e^{qt}}{e^{rt}}\right) \times [(\mu + q - r - \lambda\kappa + \widetilde{\lambda}\widetilde{\kappa})dt + \sqrt{v_t} dZ_{1,t} - \widetilde{\lambda}\widetilde{\kappa} dt + [e^J - 1]dN_t]. \tag{4.43}$$

Thus if we set the market price of Z_1 risk as

$$\zeta_{1,t} = \frac{(\mu + q - r - \lambda\kappa + \widetilde{\lambda}\widetilde{\kappa})}{\sqrt{v_t}},$$

then equation (4.43) becomes

$$d\left(\frac{S_t e^{qt}}{e^{rt}}\right) = \left(\frac{S_{t-} e^{qt}}{e^{rt}}\right) [\sqrt{v_t} d\widetilde{Z}_{1,t} - \widetilde{\lambda}\widetilde{\kappa} dt + [e^J - 1]dN_t], \tag{4.44}$$

where

$$d\widetilde{Z}_{1,t} = \zeta_{1,t} dt + dZ_{1,t},$$

with $\widetilde{Z}_{1,t}$ a standard Brownian motion under the $\mathbb{Q}$-measure.

We know that $\{\frac{C_t^E(S_t,v,t)}{e^{rt}}\}$ should be a $\mathbb{Q}$-martingale. The following steps will enable us to express this fact as a PIDE.

We note that equation (4.44) implies that

$$\frac{dS_t}{S_{t-}} = (r - q - \widetilde{\lambda}\widetilde{\kappa})dt + \sqrt{v_t} dZ_{1,t} + [e^J - 1]dN_t, \tag{4.45}$$

so that from the Feynman-Kac theorem for jumps, and the removal of discounting, the stochastic differential equation for the option price is given by

$$
\begin{aligned}
dC_t^E = \Bigg[& \frac{\partial C_{t-}^E}{\partial t} + (r - q - \widetilde{\lambda}\widetilde{\kappa})S_{t-}\frac{\partial C_{t-}^E}{\partial S} \\
& + (\kappa_v(\theta - v_t) - \lambda_v v_t)\frac{\partial C_{t-}^E}{\partial v} + \frac{v_t S_{t-}^2}{2}\frac{\partial^2 C_{t-}^E}{\partial S^2} + \rho\sigma_v v_t S_{t-}\frac{\partial^2 X_{t-}}{\partial S \partial v} \\
& + \sigma_v^2 \frac{v_t}{2}\frac{\partial^2 C_{t-}^E}{\partial v^2} + \widetilde{\lambda}\mathbb{E}_{\mathbb{Q}}^J[C_t^E(S_{t-}e^J, v, t) - C_{t-}^E]\Bigg] dt \\
& + \sqrt{v_t}S_{t-}\frac{\partial C_{t-}^E}{\partial S}d\widetilde{Z}_{1,t} + \sigma_v\sqrt{v_t}\frac{\partial C_{t-}^E}{\partial v}d\widetilde{Z}_{2,t} \\
& - \widetilde{\lambda}\mathbb{E}_{\mathbb{Q}}^J[C_t^E(S_{t-}e^J, v, t) - C_{t-}^E]dt + [C_t^E(S_{t-}e^J, v, t) - C_{t-}^E]dZ_t.
\end{aligned} \tag{4.46}
$$

In order for $\{\frac{C_t^E(S_t,v,t)}{e^{rt}}\}$ to be a $\mathbb{Q}$-martingale, the coefficient of dt in equation (4.46) in the square brackets must be equal to $rC_t^E(S_{t-}, v, t)dt$. Hence

$$
\begin{aligned}
& \frac{\partial C_{t-}^E}{\partial t} + (r - q - \widetilde{\lambda}\widetilde{\kappa})S_{t-}\frac{\partial C_{t-}^E}{\partial s} + (\kappa_v(\theta - v_t) - \lambda_v v_t)\frac{C_{t-}^E}{\partial v} \\
& \quad + \frac{v_t S_{t-}^2}{2}\frac{\partial^2 C_{t-}^E}{\partial s^2} + \rho\sigma_v v_t S_{t-}\frac{\partial^2 C_{t-}^E}{\partial s \partial v} + \sigma_v^2 \frac{v_t}{2}\frac{\partial^2 C_{t-}^E}{\partial v^2} \\
& \quad + \widetilde{\lambda}\mathbb{E}_{\mathbb{Q}}^J[C_t^E(S_{t-}e^J, v, t) - C_{t-}^E] = rC_{t-}^E.
\end{aligned} \tag{4.47}
$$

For a European style call option with final payoff $C_T^E(S_T, v_T) = (S_T - K)^+$, the option price $C_t^E(S_t, v, t)$ satisfies

$$
C_t^E(S_t, v, t) = \mathbb{E}_{\mathbb{Q}}[(S_T - K)^+ e^{-r(T-t)}|\mathcal{F}_t]. \tag{4.48}
$$

In the case of an American style call option, the option price is

$$
C_t^A(S_t, v, t) = \sup_{\tau \in [t,T]} \mathbb{E}_{\mathbb{Q}}[(S_\tau - K)^+ e^{-r(T-t)}|\mathcal{F}_t]. \tag{4.49}
$$

The following proposition gives a decomposition of the American option price in terms of its European counterpart and an early exercise premium.

Proposition 4.12. *For an American style option, the decomposition (note that we write $C^A(S, v, t)$ rather than $C^A(S, v, \tau)$ for convenience-strictly we should use a different notation for C)*

$$
C_t^A(S_t, v, t) = C_t^E(S_t, v, t) + V_t(S_t, v, t), \tag{4.50}
$$

where the first term on the right hand side is the European option price and the second term is the early exercise premium. The early exercise premium term can be written as (note that t is current time)

$$
\begin{aligned}
V_t(S_t, v, t) = &\int_t^T e^{-r\tau}\mathbb{E}_{\mathbb{Q}}[(qS_\tau - rK)1_{\{S_\tau > A_\tau\}}|A_t]d\tau \\
&- \widetilde{\lambda}\int_t^T e^{-r\tau} \\
&\times \mathbb{E}_{\mathbb{Q}}[\mathbb{E}^J_{\mathbb{Q}}[C^A_T(S_{\tau^-}e^J, v_\tau) - (S_{\tau^-}e^J - K)\mathbb{1}_{B(\tau)}]|A_\tau]d\tau,
\end{aligned}
\tag{4.51}
$$

where the event

$$B(\tau) = \{a_\tau < S_{\tau^-} \leq a_\tau e^{-J}\}$$

and where a_τ is the early exercise boundary.

Proof. For any option, whether American or European, let the discounted option prices be

$$\widetilde{C}^E_t = \frac{C^E_t(S_t, v, t)}{e^{rt}}, \quad \text{and} \quad \widetilde{C}^A_t = \frac{C^A_t(S_t, v, t)}{e^{rt}}.$$

For the discounted option prices evaluated at the pre-jump stock prices, denote

$$\widetilde{C}^E_{t-} = \frac{C^E_t(S_{t-}, v, t)}{e^{rt}}, \quad \text{and} \quad \widetilde{C}^A_{t-} = \frac{C^A_t(S_{t-}, v, t)}{e^{rt}}.$$

Applying Ito's Lemma for jump-diffusions to $\widetilde{C}^A_t$, we obtain

$$
\begin{aligned}
d\widetilde{C}^A_t = &\left[\frac{\partial^2 \widetilde{C}^A_{t-}}{\partial t} + (r - q - \tilde{\lambda}\tilde{\kappa})S_{t-}\frac{\partial \widetilde{C}^A_{t-}}{\partial S}\right. \\
&+ (\kappa_v(\theta - v_t) - \lambda_v v_t)\frac{\partial \widetilde{C}^A_{t-}}{\partial v} + \frac{1}{2}v_t S^2_{t-}\frac{\partial^2 \widetilde{C}^A_{t-}}{\partial S^2} \\
&\left. + \rho\sigma_v v_t S_{t-}\frac{\partial^2 \widetilde{C}^A_{t-}}{\partial S \partial v} + \frac{1}{2}\sigma^2_v v_t \frac{\partial^2 \widetilde{C}^A_{t-}}{\partial v^2}\right] dt \\
&+ \sqrt{v_t}S_{t-}\frac{\partial \widetilde{C}^A_{t-}}{\partial S}d\widetilde{W}_{1,t} + \sigma_v\sqrt{v_t}\frac{\partial \widetilde{C}^A_{t-}}{\partial v}d\widetilde{W}_{2,t} \\
&+ [\widetilde{C}^A_t(S_{t-}e^J, v, t) - \widetilde{C}_{t-}]dN_t.
\end{aligned}
\tag{4.52}
$$

Integrating equation (4.52) from t to T, we obtain

$$
\begin{aligned}
\frac{C^A_T}{e^{r\tau}} = &C^A_t + \int_t^T e^{-r\tau}\hat{\mathcal{L}}C^A_{\tau-}d\tau + \int_t^T e^{-r\tau}\frac{\partial C^A_{\tau-}}{\partial S}\sqrt{v_\tau}S_{\tau-}d\widetilde{W}_{1,\tau} \\
&+ \int_t^T e^{-r\tau}\frac{\partial C^A_{\tau-}}{\partial v}\sigma_v\sqrt{v_\tau}d\widetilde{W}_{2,\tau}
\end{aligned}
$$

$$+e^{-r\tau}\left[-\tilde{\lambda}\int_t^T \mathbb{E}^J_{\mathbb{Q}}[C^A_\tau(S_{\tau-}e^J, v, t) - C^A_{\tau-}]d\tau\right.$$

$$\left.+\int_t^T [C^A_\tau(S_{\tau-}e^J, v, t) - C^A_{\tau-}]dN_\tau\right], \tag{4.53}$$

where

$$\begin{aligned}\hat{\mathcal{L}}C^A_{\tau-} &= \frac{\partial C^A_{\tau-}}{\partial t} - rC^A_{\tau-} + (r - q - \tilde{\lambda}\tilde{\kappa})S_{\tau-}\frac{\partial C^A_\tau}{\partial S}\\ &+(\kappa_v(\theta - v_t) - \lambda_v v_t)\frac{\partial C^A_{t-}}{\partial v}\\ &+\frac{1}{2}v_t S^2_{\tau-}\frac{\partial^2 C^A_{\tau-}}{\partial S^2} + \rho\sigma_v v_t S_{t-}\frac{\partial^2 C^A_{t-}}{\partial S\partial v} + \frac{1}{2}\sigma_v^2 v_t\frac{\partial^2 C^A_{t-}}{\partial v^2}\\ &+\tilde{\lambda}\mathbb{E}^J_{\mathbb{Q}}[C^A_\tau(S_{\tau-}e^J, v, t) - C^A_{\tau-}].\end{aligned} \tag{4.54}$$

Now take conditional expectation of equation (4.53) under the martingale measure $\mathbb{Q}$ and conditioning on $\mathcal{F}_t$, the discounted conditional expectation of the final payoff of the American option is

$$\mathbb{E}_{\mathbb{Q}}\left[\frac{C^A_T}{e^{r(T-t)}}\middle|\mathcal{F}_t\right] = C^A_t + \mathbb{E}_{\mathbb{Q}}\left[\int_t^T e^{-r\tau}\hat{\mathcal{L}}C^A_{\tau-}d\tau\middle|\mathcal{F}_t\right]. \tag{4.55}$$

Note that the conditional expectation of the other remaining terms in equation (4.53) are zero since they are all (local) martingales with zero mean.

At maturity time T, the final payoffs of the American and European calls are the same, that is, $C^A_T(S_T, v_T) = C^E_T(S_T, v_T) = (S_T - K)^+$. Hence equation (4.55) simplifies to

$$C^A_t = C^E_t - \int_t^T e^{-r\tau}\mathbb{E}_{\mathbb{Q}}[\hat{\mathcal{L}}C^A_{\tau-} \mid \mathcal{F}_t]d\tau. \tag{4.56}$$

The boundary and smooth-pasting conditions given the early exercise boundary a_τ are

$$\begin{aligned}\lim_{S\to a_\tau}\frac{\partial C^A_\tau(S)}{\partial S} &= 1,\\ C^A_\tau a(v,\tau) &= a(v,\tau) - K,\\ C^A_\tau(S_\tau) &= S_\tau - K \quad \text{where } S_\tau > a_\tau,\\ \frac{\partial C^A_\tau(S)}{\partial t} &= 0, \quad \text{when } S > a_\tau \quad \text{and}\\ \frac{\partial C^A_\tau(S)}{\partial v} &= 0, \quad \text{when } S > a_\tau.\end{aligned}$$

In equation (4.56), the partial-integro differential operator

$$\hat{\mathcal{L}}C^A_{\tau-} < 0$$

is in the early exercise region $S_{\tau-} > a(v,\tau)$, since the American option is a strict supermartingale in that region. In the continuation region $S_{\tau-} \leq a(v,\tau)$, it is not optimal to exercise the American option and hence it behaves like a European option. Thus

$$\hat{\mathcal{L}}C^A_{\tau-} = 0.$$

Denote the early exercise by

$$\mathcal{A} = \{S_{\tau-} > a(v,\tau)\}.$$

Hence

$$\begin{aligned}
C^A_t &= C^E_t - \int_t^T e^{-r\tau}\mathbb{E}_{\mathbb{Q}}[\hat{\mathcal{L}}C^A_{\tau-}\mathbb{1}_{\{S_{\tau-}>a(v,\tau)\}} + \hat{\mathcal{L}}C^A_{\tau-}\mathbb{1}_{\{S_{\tau-}>a(v,\tau)\}^C} \mid \mathcal{F}_t]d\tau \\
&= C^E_t - \int_t^T e^{-r\tau}\mathbb{E}_{\mathbb{Q}}[\hat{\mathcal{L}}C^A_{\tau-}\mathbb{1}_{\{S_{\tau-}>a(v,\tau)\}} \mid \mathcal{F}_t]d\tau \\
&= C^E_t - \int_t^T e^{-r\tau} \\
&\quad \times\mathbb{E}_{\mathbb{Q}}[(-r(S_{\tau-}-K) + (r-q-\tilde{\lambda}\tilde{\kappa})S_{\tau-})\mathbb{1}_{\{S_{\tau-}>a(v,\tau)\}} \mid \mathcal{F}_t]d\tau \\
&\quad +\tilde{\lambda}\int_t^T e^{-r\tau} \\
&\quad \times E_{\mathbb{Q}}[\mathbb{E}^J_{\mathbb{Q}}(C^A_\tau(S_\tau - e^J) - (S_{\tau-}-K))\mathbb{1}_{\{S_{\tau-}>a(v,\tau)\}} \mid \mathcal{F}_t]d\tau \\
&= C^E_t - \int_t^T e^{-r\tau}\mathbb{E}_{\mathbb{Q}}[(qS_{\tau-} - rK)\mathbb{1}_{\{S_{\tau-}>a(v,\tau)\}} \mid \mathcal{F}_t]d\tau \\
&\quad - \int_t^T e^{-r\tau} \\
&\quad \times\mathbb{E}_{\mathbb{Q}}[\mathbb{E}^J_{\mathbb{Q}}[C^A_\tau(S_{\tau-}e^J, v, t) - (S_{\tau-}e^J - K)]\mathbb{1}_{\{S_{\tau-}>a(v,\tau)\}} \mid \mathcal{F}_t]d\tau.
\end{aligned} \tag{4.57}$$

Note that in equation (4.57)

$$C^A_\tau(S_{\tau-}e^J) - (S_{\tau-}e^J - K) = 0$$

if $S_{\tau-}e^J \geq a(v,\tau)$, and

$$C^A_\tau(S_{\tau-}e^J) - (S_{\tau-}e^J - K) > 0$$

if $S_{\tau-}e^J < a(v,\tau)$. Hence equation (4.57) can be written as

$$C^A_t = C^E_t + \int_t^T e^{-r\tau}\mathbb{E}_{\mathbb{Q}}[(qS_{\tau-} - rK)\mathcal{L}_{\{S_{\tau-}>a(v,\tau)\}} \mid \mathcal{F}_t]d\tau$$

$$
- \int_t^T e^{-r\tau} \mathbb{E}_{\mathbb{Q}}[\mathbb{E}^J_{\mathbb{Q}}[C^A_\tau(S_{\tau-}e^J) - (S_{\tau-}e^J - K)]
$$

$$
\times \mathbb{1}_{\{a(v,\tau)<S_\tau<a(v,\tau)e^{-J}\}} \mid \mathcal{F}_t]d\tau.
$$

□

4.5 Conclusion

In this chapter we have studied the evaluation of American call options under stochastic volatility and jump-diffusion. The free boundary problem for the American call involves a homogeneous partial-integro differential equation (PIDE) which must be solved in a restricted domain. Using a method proposed by Jamshidian (1992), this PIDE is transformed into an equivalent inhomogenous PIDE in an unrestricted domain, which is then solved using a Fourier transform approach. The solution for the American call involves a linked system of integral equations for the option price and early exercise surface. The difficulties involved in solving this system of integral equations are also discussed.

Appendix

4.A Deriving the Inhomogeneous PIDE

To derive the inhomogeneous PIDE in the region $0 \leq S < \infty$ that corresponds to the free boundary value problem (2.12)–(2.15), we must find an inhomogeneous term such that (2.12) is true for all values of S. Noting that $C^A(S, v, \tau) = S - K$ when $S \geq a(v, \tau)$, the PIDE in the stopping region evaluates to

$$
I(S, v, \tau) = \mathbb{1}_{(S-a(v,\tau)\leq 0)} \Bigg\{ (r - q - \lambda^* k^*)S - r(S - K)
$$

$$
+ \lambda^* \int_0^\infty [C^A(SY, v, \tau) - (S - K)]G^*(Y)dY, \Bigg\},
$$

where λ^*, $G^*(Y)$ and k^* are defined by (2.17) and (2.18). By noting that the when $SY > a(v, \tau)$ we have $C^A(SY, v, \tau) = SY - K$, we can write $I(S, v, \tau)$ as

$$
I(S, v, \tau) = \mathbb{1}_{(S-a(v,\tau)\leq 0)} \Bigg\{ rK - qS - \gamma k S
$$

$$
+ \lambda^* \int_0^{a(v,\tau)/S} [C^A(SY, v, \tau) - (S - K)]G^*(Y)dY
$$

$$+\lambda^* \int_{a(v,\tau)/S}^{\infty} [(SY-K)-(S-K)]G^*(Y)dY \Bigg\}$$

$$= \mathbb{1}_{(S-a(v,\tau)\leq 0)} \Bigg\{ rK - qS - \lambda^* k^* S$$

$$+\lambda^* \int_0^{a(v,\tau)/S} [C^A(SY,v,\tau)-(S-K)]G^*(Y)dY$$

$$+\lambda^* \int_0^{\infty} S(Y-1)G^*(Y)dY$$

$$-\lambda^* \int_0^{a(v,\tau)/S} S(Y-1)G^*(Y)dY \Bigg\},$$

which, by use of the definition for k^* becomes

$$I(S,v,\tau) = \mathbb{1}_{(S-a(v,\tau)\leq 0)} \Bigg\{ rK - qS$$

$$+\lambda^* \int_0^{a(v,\tau)/S} [C^A(SY,v,\tau)-(SY-K)]G^*(Y)dY \Bigg\}.$$

Thus the inhomogeneous PIDE for the American call option price in the domain $0 \leq S < \infty$ is given by

$$\frac{\partial C^A}{\partial \tau} = \frac{vS^2}{2}\frac{\partial^2 C^A}{\partial S^2} + \rho\sigma_v vS\frac{\partial^2 C^A}{\partial S \partial v} + \frac{\sigma_v^2 v}{2}\frac{\partial^2 C^A}{\partial v^2}$$

$$+ (r-q-\lambda^* k^*)S\frac{\partial C^A}{\partial S} + (\kappa[\theta - v] - \lambda_v v)\frac{\partial C^A}{\partial v} - rC$$

$$+\lambda^* \int_0^{\infty} [C^A(SY,v,\tau) - C^A(S,v,\tau)]G^*(Y)dY + f(S,v,\tau),$$

where $f(S,v,\tau) = -I(S,v,\tau)$.

4.B Verifying Duhamel's Principle

Consider the PIDE

$$\frac{\partial C^A}{\partial \tau} = \mathcal{D}_{S,v}C^A - rC^A + f(S,v,\tau) \tag{4.58}$$

with initial condition $C^A(S,v,0) = c(S,v)$, to be solved in the region $\tau \geq 0$, $0 \leq S < \infty$ and $0 \leq v < \infty$, and where we define the integro-differential operator $\mathcal{D}_{S,v}$ as

$$\mathcal{D}_{S,v}C^A(S,v) \equiv \frac{vS^2}{2}\frac{\partial^2 C^A}{\partial S^2} + \rho\sigma_v vS\frac{\partial^2 C^A}{\partial S\partial v} + \frac{\sigma_v^2 v}{2}\frac{\partial^2 C^A}{\partial v^2}$$
$$+ (r - q - \lambda^* k^*)S\frac{\partial C^A}{\partial S} + (\kappa[\theta - v] - \lambda_v v)\frac{\partial C^A}{\partial v}$$
$$+ \lambda^* \int_0^\infty [C^A(SY, v, \tau) - C^A(S, v, \tau)]G^*(Y)dY.$$

In Sec. 5.3 we show that when $f(S, v, \tau) = 0$,

$$C_E(S, v, \tau) = e^{-r\tau}\int_0^\infty \int_{-\infty}^\infty c(e^y, w)\bar{G}(y, w, \tau; S, v)dydw$$

is the solution to (4.58), where the Green's function[6] $\bar{G}(y, w, \tau; S, v)$ satisfies the PIDE

$$\frac{\partial \bar{G}}{\partial \tau} = \mathcal{D}_{S,v}\bar{G}.$$

The initial condition is $\bar{G}(y, w, 0; S, v) = \delta(w - v)\delta(y - \ln S)$, where $\delta(x)$ is the Dirac delta function.

According to Duhamel's principle, the solution to the PIDE (4.58) when $f(S, v, \tau) \neq 0$ is given by

$$C^A(S, v, \tau) = e^{-r\tau}\int_0^\infty \int_{-\infty}^\infty c(e^y, w)\bar{G}(y, w, \tau; S, v)dydw$$
$$+ \int_0^\tau e^{-r(\tau-\xi)}\int_0^\infty \int_{-\infty}^\infty f(e^y, w, \xi)\bar{G}(y, w, \tau - \xi; S, v)dydwd\xi$$
$$\equiv C_E(S, v, \tau) + C_P(S, v, \tau). \tag{4.59}$$

We can readily verify that this is the correct solution by showing that (4.59) satisfies the PIDE (4.58). Substituting $C^A(S, v, \tau) = C_E(S, v, \tau) + C_P(S, v, \tau)$ into (4.58), we have

$$\frac{\partial C^A}{\partial \tau} + rC - \mathcal{D}_{s,v}C^A - f(s, v, \tau)$$
$$= e^{-r\tau}\int_0^\infty \int_{-\infty}^\infty c(e^y, w)\left\{\frac{\partial \bar{G}}{\partial \tau} - \mathcal{D}_{S,v}\bar{G}\right\}dydw - rC_E + rC_E$$

[6]For more on the theory of Green's functions, we refer the reader to Duffy (2001).

$$
\begin{aligned}
&+\int_0^\infty \int_{-\infty}^\infty f(e^y, w, \tau)\bar{G}(y, w, 0; S, v)dydw \\
&+\int_0^\tau e^{-r(\tau-\xi)} \int_0^\infty \int_{-\infty}^\infty f(e^y, w, \xi)\frac{\partial \bar{G}}{\partial \tau} dydwd\xi - rC_P + rC_P \\
&-\int_0^\tau e^{-r(\tau-\xi)} \int_0^\infty \int_{-\infty}^\infty f(e^y, w, \xi)\mathcal{D}_{S,v}\bar{G}dydwd\xi - f(s, v, \tau) \\
&= \int_0^\infty \int_{-\infty}^\infty f(e^y, w, \tau)\delta(y - \ln S)\delta(w - v)dydw \\
&+\int_0^\tau e^{-r(\tau-\xi)} \int_0^\infty \int_{-\infty}^\infty f(e^y, w, \xi) \\
&\times \left[\frac{\partial \bar{G}}{\partial \tau} - \mathcal{D}_{s,v}\bar{G}\right] dydwd\xi - f(S, v, \tau) \\
&= f(S, v, \tau) + 0 - f(s, v, \tau) = 0.
\end{aligned}
$$

Hence $C^A(S, v, \tau)$ satisfies the PIDE (4.58).

4.C Proof of Proposition 4.3 — Fourier Transform of the PIDE

By use of (4.13), we can readily show that

$$
\mathcal{F}_x\left\{\frac{\partial U}{\partial x}\right\} = -i\phi\hat{U}, \quad \mathcal{F}_x\left\{\frac{\partial^2 U}{\partial x^2}\right\} = -\phi^2\hat{U},
$$

$$
\mathcal{F}_x\left\{\frac{\partial^2 U}{\partial x \partial v}\right\} = -i\phi\frac{\partial \hat{U}}{\partial v}, \quad \mathcal{F}_x\left\{\frac{\partial U}{\partial \tau}\right\} = \frac{\partial \hat{U}}{\partial \tau}.
$$

All that remains is to evaluate the transform of the integral term. Applying the definition of the transform (4.13), we have

$$
\begin{aligned}
&F_x\left\{\int_0^\infty U(x + \ln Y, v, \tau)G^*(Y)dY\right\} \\
&\quad = \int_0^\infty \int_{-\infty}^\infty e^{i\phi x}U(x + \ln Y, v, \tau)G^*(Y)dxdY.
\end{aligned}
$$

Making the change of integration variable $z = x + \ln Y$ this becomes

$$
F_x\left\{\int_0^\infty U(x + \ln Y, v, \tau)G^*(Y)dY\right\}
$$

$$= \int_0^\infty G^*(Y) \int_{-\infty}^\infty e^{i\phi(z-\ln Y)} U(z,v,\tau) dz dY$$

$$= \int_0^\infty G^*(Y) e^{-i\phi \ln Y} dY \int_{-\infty}^\infty e^{i\phi z} U(z,v,\tau) dz$$

$$= A(\phi)\hat{U}(\phi,v,\tau),$$

where $A(\phi)$ is defined in (4.15). Simple factorisation then yields the two-dimensional PDE (4.14) in Proposition 4.3.

4.D Proof of Proposition 4.4 — Laplace Transform of the PDE (4.14)

Taking the Laplace transform (4.16) of (4.14), we require the following results. Firstly,

$$\mathcal{L}_v \left\{ \frac{\Lambda v}{2} \hat{U} - i\phi\Psi\hat{U} \right\} = \frac{\Lambda}{2} \int_0^\infty v e^{-sv} \hat{U} dv - i\phi\Psi\tilde{U}$$

$$= -\frac{\Lambda}{2} \frac{\partial}{\partial s} \int_0^\infty e^{-sv} \hat{U} dv - i\phi\Psi\tilde{U}$$

$$= -\frac{\Lambda}{2} \frac{\partial \tilde{U}}{\partial s} - i\phi\Psi\tilde{U}.$$

For the first order derivative with respect to v we have

$$\mathcal{L}_v \left\{ (\alpha - \Theta v) \frac{\partial \hat{U}}{\partial v} \right\} = \int_0^\infty (\alpha - \Theta v) e^{-sv} \frac{\partial \hat{U}}{\partial v} dv$$

$$= \alpha \left\{ -\hat{U}(\phi,0,\tau) + s \int_0^\infty e^{-sv} \hat{U} dv \right\}$$

$$+ \Theta \frac{\partial}{\partial s} \int_0^\infty e^{-sv} \frac{\partial \hat{U}}{\partial v} dv$$

$$= \alpha(-\hat{U}(\phi,0,\tau) + s\tilde{U}) + \Theta \frac{\partial}{\partial s} \{-\hat{U}(\phi,0,\tau) + s\tilde{U}\}$$

$$= -\alpha\hat{U}(\phi,0,\tau) + \alpha s\tilde{U} + \Theta \left(s \frac{\partial \tilde{U}}{\partial s} + \tilde{U} \right)$$

$$= \Theta s \frac{\partial \tilde{U}}{\partial s} + (\alpha s + \Theta)\tilde{U} - \alpha\hat{U}(\phi,0,\tau).$$

Finally, for the second order derivative term,

$$
\begin{aligned}
\mathcal{L}_v\left\{\frac{\sigma_v^2 v}{2}\frac{\partial^2 \hat{U}}{\partial v^2}\right\} &= -\frac{\sigma_v^2}{2}\frac{\partial}{\partial s}\int_0^\infty e^{-sv}\frac{\partial^2 \hat{U}}{\partial v^2}dv \\
&= -\frac{\sigma_v^2}{2}\frac{\partial}{\partial s}\left\{s\int_0^\infty e^{-sv}\frac{\partial \hat{U}}{\partial v}dv\right\} \\
&= -\frac{\sigma_v^2}{2}\frac{\partial}{\partial s}\left\{s[-\hat{U}(\phi,0,\tau)] + s^2\tilde{U}\right\} \\
&= -\frac{\sigma_v^2}{2}\left(-\hat{U}(\phi,0,\tau) + 2s\tilde{U} + s^2\frac{\partial \tilde{U}}{\partial s}\right) \\
&= \frac{-\sigma_v^2 s^2}{2}\frac{\partial \tilde{U}}{\partial s} - \sigma_v^2 s\tilde{U} + \frac{\sigma_v^2}{2}\hat{U}(\phi,0,\tau).
\end{aligned}
$$

Thus the Laplace transform of (4.14) is

$$
\begin{aligned}
&\frac{\partial \tilde{U}}{\partial \tau} + \left(\frac{\sigma_v^2}{2}s^2 - \Theta s + \frac{\Lambda}{2}\right)\frac{\partial \tilde{U}}{\partial s} \\
&\quad = [(\alpha - \sigma_v^2)s + \Theta - i\phi\Psi]\tilde{U} + \left(\frac{\sigma_v^2}{2} - \alpha\right)\hat{U}(\phi,0,\tau).
\end{aligned}
$$

Finally we set $f_L(\tau) = (\sigma_v^2/2 - \alpha)\hat{U}(\phi,0,\tau)$, and note that since $\tilde{U}(\phi,s,\tau)$ must be finite for all $s > 0$, $f_L(\tau)$ must be determined such that $\tilde{U}(\phi,s,\tau) \to 0$ as $s \to \infty$.

4.E Proof of Proposition 4.5 — Solving the PDE (4.17)

First we express the solution in terms of the so far unknown function $f_L(t)$. Since (4.17) is a first order PDE, it may be solved using the method of characteristics. The characteristic equation for (4.17) is

$$
d\tau = \frac{ds}{\left(\frac{\sigma_v^2}{2}s^2 - \Theta s + \frac{\Lambda}{2}\right)} = \frac{d\tilde{U}}{[(\alpha - \sigma_v^2)s + \Theta - i\phi\Psi]\tilde{U} + f_L(\tau)}. \tag{4.60}
$$

Taking the first pair in (4.60), we have

$$
\int d\tau = \frac{2}{\sigma_v^2}\int \frac{ds}{s^2 - \frac{2\Theta}{\sigma_v^2}s + \frac{\Lambda}{\sigma_v^2}}
$$

so that[7]

$$\tau + c_1 = \frac{1}{\Omega}\int\left(\frac{1}{s - (\frac{\Gamma+\Omega}{\sigma_v^2})} - \frac{1}{s - (\frac{\Gamma-\Omega}{\sigma_v^2})}\right) ds,$$

where we use the notation c_j to denote an undetermined constant term. Integrating with respect to s gives a relation between the transform variable s and time-to-maturity τ, namely

$$\Omega\tau + c_2 = \ln\left(\frac{\sigma_v^2 s - \Theta - \Omega}{\sigma_v^2 s - \Theta + \Omega}\right),$$

and hence we have

$$c_3 = \frac{(\sigma_v^2 s - \Theta - \Omega)e^{-\Omega\tau}}{\sigma_v^2 s - \Theta + \Omega}. \tag{4.61}$$

We also note that (4.61) may be re-expressed as

$$s = \frac{(\Theta - \Omega)}{\sigma_v^2} - \frac{2\Omega e^{-\Omega\tau}}{\sigma_v^2(c_3 - e^{-\Omega\tau})}. \tag{4.62}$$

We next consider the last pair in (4.60), which can be rearranged to give the first order ODE

$$\frac{d\tilde{U}}{d\tau} + [(\sigma_v^2 - \alpha)s - \Theta + i\phi\Psi]\tilde{U} = f_L(\tau). \tag{4.63}$$

The integrating factor, $R(\tau)$, for this ODE is the solution to

$$\frac{dR}{d\tau} = [(\sigma_v^2 - \alpha)s - \Theta + i\phi\Psi]R.$$

Using the expression (4.62) for s and integrating with respect to τ gives

$$\ln R = \left[\frac{(\sigma_v^2 - \alpha)(\Theta - \Omega)}{\sigma_v^2} - \Theta + i\phi\Psi\right]\tau - (\sigma_v^2 - \alpha)\int\frac{2\Omega e^{-\Omega\xi}}{\sigma_v^2(c_3 - e^{-\Omega\xi})}d\xi.$$

Using the change of integration variable $u = c_3 - e^{-\Omega\xi}$, we have

$$\int\frac{e^{-\Omega\xi}}{c_3 - e^{-\Omega\xi}}d\xi = \frac{1}{\Omega}\ln|u|,$$

and hence

$$R(\tau) = \exp\left\{\left[\frac{(\sigma_v^2 - \alpha)(\Theta - \Omega)}{\sigma_v^2} - \Theta + i\phi\Psi\right]\tau\right\}\left|\frac{1}{c_3 - e^{-\Omega\tau}}\right|^{\frac{2}{\sigma_v^2}(\sigma_v^2 - \alpha)}. \tag{4.64}$$

---

[7]Note that $x^2 - \frac{2\Theta}{\sigma_v^2}x + \frac{\Lambda}{\sigma_v^2} = 0$ has solution $x = \frac{\Theta\pm\Omega}{\sigma_v^2}$ where we define $\Omega = \Omega(\phi) \equiv \sqrt{\Theta^2 - \Lambda\sigma_v^2}$.

Thus applying the method of variation of parameters to solve the ordinary differential equation (4.63) we find that $\tilde{U}(\phi, s, \tau)$ is given by

$$R(\tau)\tilde{U}(\phi, s, \tau) = \int_0^\tau R(\xi) f_L(\xi) d\xi + c_4,$$

which on use of the expression equation (4.64) for $R(\tau)$ becomes

$$\tilde{U}(\phi, s, \tau) = \exp\left\{\left[\frac{(\alpha - \sigma_v^2)(\Theta - \Omega)}{\sigma_v^2} + \Theta - i\phi\Psi\right]\tau\right\} |c_3 - e^{-\Omega\tau}|^{\frac{2}{\sigma_v^2}(\sigma_v^2 - \alpha)}$$
$$\times \left\{c_4 + \int_0^\tau f_L(\xi) \exp\left\{\left[\frac{(\sigma_v^2 - \alpha)(\Theta - \Omega)}{\sigma_v^2} - \Theta + i\phi\Psi\right]\xi\right\}\right.$$
$$\left.\times \left|\frac{1}{c_3 - e^{-\Omega\xi}}\right|^{\frac{2}{\sigma_v^2}(\sigma_v^2 - \alpha)} d\xi\right\}. \tag{4.65}$$

Next we determine the constant c_4 that appears in equation (4.65). We anticipate that we will find a function A such that $c_4 = A(c_3)$, where c_3 is given by (4.61). When $\tau = 0$, we have from equation (4.61) and equation (4.65) that

$$\tilde{U}(\phi, s, 0) = \left|\frac{\sigma_v^2 s - \Theta - \Omega}{\sigma_v^2 s - \Theta + \Omega} - 1\right|^{\frac{2}{\sigma_v^2}(\sigma_v^2 - \alpha)} A(c_3). \tag{4.66}$$

Note from (4.62) that at $\tau = 0$ we have

$$s = \frac{\Theta - \Omega}{\sigma_v^2} - \frac{2\Omega}{(c_3 - 1)\sigma_v^2}. \tag{4.67}$$

Thus from (4.66) we find that $A(c_3)$ is given by

$$A(c_3) = |c_3 - 1|^{-\frac{2}{\sigma_v^2}(\sigma_v^2 - \alpha)} \tilde{U}\left(\phi, \frac{\Theta - \Omega}{\sigma_v^2} - \frac{2\Omega}{(c_3 - 1)\sigma_v^2}, 0\right).$$

In equation (4.65) consider the term

$$|c_3 - e^{-\Omega\tau}|^{\frac{2}{\sigma_v^2}(\sigma_v^2 - \alpha)} A(c_3).$$
$$= \left|\frac{2\Omega e^{-\Omega\tau}}{(\sigma_v^2 s - \Theta + \Omega)(1 - e^{-\Omega\tau}) + 2\Omega e^{-\Omega\tau}}\right|^{\frac{2}{\sigma_v^2}(\sigma_v^2 - \alpha)} \tilde{U}$$
$$\times \left(\phi, \frac{\Theta - \Omega}{\sigma_v^2} - \frac{2\Omega}{(c_3 - 1)\sigma_v^2}, 0\right),$$

and note from (4.61) that

$$\frac{2\Omega}{(c_3 - 1)\sigma_v^2} = \frac{(\sigma_v^2 s - \Gamma + \Omega)2\Omega}{\sigma_v^2[(e^{-\Omega\tau} - 1)(\sigma_v^2 s - \tau + \Omega) - 2\Omega e^{-\Omega\tau}]}.$$

Also, we note that for $0 \leq \xi \leq \tau$, we have

$$\left|\frac{c_3 - e^{-\Omega\tau}}{c_3 - e^{-\Omega t}}\right|^{\frac{2}{\sigma_v^2}(\sigma_v^2-\alpha)}$$

$$= \left|\frac{(\sigma_v^2 s - \Theta - \Omega)e^{-\Omega\tau} - (\sigma_v^2 s - \Theta + \Omega)e^{-\Omega\tau}}{(\sigma_v^2 s - \Theta - \Omega)e^{-\Omega\tau} - (\sigma_v^2 s - \Theta + \Omega)e^{-\Omega t}}\right|^{\frac{2}{\sigma_v^2}(\sigma_v^2-\alpha)}$$

$$= \left|\frac{2\Omega e^{-\Omega\tau}}{(\sigma_v^2 s - \Theta + \Omega)(e^{-\Omega t} - e^{-\Omega\tau}) + 2\Omega e^{-\Omega\tau}}\right|^{\frac{2}{\sigma_v^2}(\sigma_v^2-\alpha)},$$

and make the observation that all real arguments in $|\cdot|$ are positive. Thus by substituting the last equation into (4.65) we have

$$\begin{aligned}
\tilde{U}(\phi, s, \tau) &= \exp\left\{\left[\frac{(\alpha - \sigma_v^2)(\Theta - \Omega)}{\sigma_v^2} + \Theta - i\phi\Psi\right]\tau\right\} \\
&\quad \times \tilde{U}\left(\phi, \frac{\Theta - \Omega}{\sigma_v^2} - \frac{(\sigma_v^2 s - \Theta + \Omega)2\Omega}{\sigma_v^2[(e^{-\Omega\tau} - 1)(\sigma_v^2 s - \tau + \Omega) - 2\Omega e^{-\Omega\tau}]}, 0\right) \\
&\quad \times \left(\frac{2\Omega e^{-\Omega\tau}}{(\sigma_v^2 s - \Theta + \Omega)(1 - e^{-\Omega\tau}) + 2\Omega e^{-\Omega\tau}}\right)^{\frac{2}{\sigma_v^2}(\sigma_v^2-\alpha)} \\
&\quad + \int_0^\tau f_L(\xi) \exp\left\{\left[\frac{(\alpha - \sigma_v^2)(\Theta - \Omega)}{\sigma_v^2} + \Theta - i\phi\Psi\right](\tau - \xi)\right\} \\
&\quad \times \left(\frac{2\Omega e^{-\Omega\tau}}{(\sigma_v^2 s - \Theta + \Omega)(e^{-\Omega\xi} - e^{-\Omega\tau}) + 2\Omega e^{-\Omega\tau}}\right)^{\frac{2}{\sigma_v^2}(\sigma_v^2-\alpha)} d\xi.
\end{aligned} \tag{4.68}$$

We must now determine the function $f_L(\xi)$: We achieve this by applying the condition (4.18) to (4.68). Taking the limit of (4.68) as $s \to \infty$, we require that

$$\begin{aligned}
&\int_0^\tau f_L(\xi) \exp\left\{-\left[\frac{(\alpha - \sigma_v^2)(\Theta - \Omega)}{\sigma_v^2} + \Theta - i\phi\Psi\right]\xi\right\} \\
&\qquad \times \left(\frac{1 - e^{-\Omega\tau}}{e^{-\Omega\xi} - e^{-\Omega\tau}}\right)^{\frac{2}{\sigma_v^2}(\sigma_v^2-\alpha)} d\xi \\
&\quad = -\tilde{U}\left(\phi, \frac{\Theta - \Omega}{\sigma_v^2} - \frac{2\Omega}{\sigma_v^2(e^{-\Omega\tau} - 1)}, 0\right).
\end{aligned} \tag{4.69}$$

In equation (4.69) make the change of variable

$$\zeta^{-1} = 1 - e^{-\Omega\xi}, \qquad z^{-1} = 1 - e^{-\Omega\tau}, \tag{4.70}$$

so that

$$\int_z^\infty g(\zeta)(\zeta - z)^{\frac{2}{\sigma_v^2}(\alpha-\sigma_v^2)} d\zeta = -\Omega\tilde{U}\left(\phi, \frac{\Theta - \Omega}{\sigma_v^2}, +\frac{2\Omega z}{\sigma_v^2}, 0\right), \tag{4.71}$$

where

$$g(\zeta) = f_L(\xi) \exp\left\{-\left[\frac{(\alpha - \sigma_v^2)(\Theta - \Omega)}{\sigma_v^2} + \Theta - i\phi\Psi\right]\xi\right\} \frac{\zeta^{\frac{2}{\sigma_v^2}(\sigma_v^2 - \alpha)}}{\zeta(\zeta - 1)}. \tag{4.72}$$

Thus our task is to solve (4.71) for $g(\zeta)$, and hence we will obtain the function $f_L(\xi)$.

Firstly, by definition (4.16) for the Laplace transform,

$$\begin{aligned}
&\tilde{U}\left(\phi, \frac{\Theta - \Omega}{\sigma_v^2} + \frac{2\Omega z}{\sigma_v^2}, 0\right) \\
&\quad = \int_0^\infty \hat{U}(\phi, w, 0) \exp\left\{-\left(\frac{\Theta - \Omega}{\sigma_v^2} + \frac{2\Omega z}{\sigma_v^2}\right) w\right\} dw.
\end{aligned}$$

Introducing a gamma function, as defined by (4.22), we have

$$\begin{aligned}
&\tilde{U}\left(\phi, \frac{\Theta - \Omega}{\sigma_v^2} + \frac{2\Omega z}{\sigma_v^2}, 0\right) \\
&= \frac{\Gamma\left(\frac{2\alpha}{\sigma_v^2} - 1\right)}{\Gamma\left(\frac{2\alpha}{\sigma_v^2} - 1\right)} \int_0^\infty \hat{U}(\phi, w, 0) \exp\left\{-\left(\frac{\Theta - \Omega}{\sigma_v^2} + \frac{2\Omega z}{\sigma_v^2}\right) w\right\} dw \\
&= \frac{1}{\Gamma\left(\frac{2\alpha}{\sigma_v^2} - 1\right)} \int_0^\infty \int_0^\infty e^{-a} a^{\frac{2\alpha}{\sigma_v^2} - 2} \hat{U}(\phi, w, 0) \\
&\quad \times \exp\left\{-\left(\frac{\Theta - \Omega}{\sigma_v^2} + \frac{2\Omega z}{\sigma_v^2}\right) w\right\} da dw.
\end{aligned}$$

The change of integration variable $a = 2\Omega w/\sigma_v^2 y$ gives

$$\begin{aligned}
&\tilde{U}\left(\phi, \frac{\Theta - \Omega}{\sigma_v^2} + \frac{2\Omega z}{\sigma_v^2}, 0\right) \\
&= \frac{1}{\Gamma\left(\frac{2\alpha}{\sigma_v^2} - 1\right)} \int_0^\infty \hat{U}(\phi, w, 0) \exp\left\{-\left(\frac{\Theta - \Omega}{\sigma_v^2}\right) w\right\} \\
&\quad \times \left[\int_0^\infty e^{-\frac{2\Omega w}{\sigma_v^2} y} \left(\frac{2\Omega w}{\sigma_v^2} y\right)^{\frac{2\alpha}{\sigma_v^2} - 2} e^{-\frac{2\Omega z}{\sigma_v^2} w} dy\right] \left(\frac{2\Omega w}{\sigma_v^2}\right) dw
\end{aligned}$$

$$= \frac{1}{\Gamma\left(\frac{2\alpha}{\sigma_v^2}-1\right)} \int_0^\infty \hat{U}(\phi, w, 0) \exp\left\{-\left(\frac{\Theta-\Omega}{\sigma_v^2}\right) w\right\} \left(\frac{2\Omega w}{\sigma_v^2}\right)^{\frac{2\alpha}{\sigma_v^2}-1}$$
$$\times \left(\int_0^\infty e^{-\frac{2\Omega w}{\sigma_v^2}(y+z)} y^{\frac{2\alpha}{\sigma_v^2}-2} dy\right) dw.$$

Making one further change of variable, namely $\zeta = z + y$, we have

$$\tilde{U}\left(\phi, \frac{\Theta-\Omega}{\sigma_v^2} + \frac{2\Omega z}{\sigma_v^2}, 0\right)$$
$$= \frac{1}{\Gamma\left(\frac{2\alpha}{\sigma_v^2}-1\right)} \int_0^\infty \hat{U}(\phi, w, 0) \exp\left\{-\left(\frac{\Theta-\Omega}{\sigma_v^2}\right) w\right\} \left(\frac{2\Omega w}{\sigma_v^2}\right)^{\frac{2\alpha}{\sigma_v^2}-1}$$
$$\times \int_z^\infty e^{-\frac{2\Omega w}{\sigma_v^2}\zeta} (\zeta - z)^{\frac{2\alpha}{\sigma_v^2}-2} d\zeta dw$$
$$= \int_z^\infty (\zeta - z)^{\frac{2}{\sigma_v^2}(\alpha - \sigma_v^2)} \left[\int_0^\infty \frac{\hat{U}(\phi, w, 0)}{\Gamma\left(\frac{2\alpha}{\sigma_v^2}-1\right)} \left(\frac{2\Omega w}{\sigma_v^2}\right)^{\frac{2\alpha}{\sigma_v^2}-1}\right.$$
$$\left.\times \exp\left\{-\left(\frac{\Theta-\Omega}{\sigma_v^2} + \frac{2\Omega\zeta}{\sigma_v^2}\right) w\right\} dw\right] d\zeta. \tag{4.73}$$

Comparing (4.73) with (4.71), we can conclude that

$$g(z) = \frac{-\Omega}{\Gamma\left(\frac{2\alpha}{\sigma_v^2}-1\right)} \int_0^\infty \hat{U}(\phi, w, 0) \left(\frac{2\Omega u}{\sigma_v^2}\right)^{\frac{2\alpha}{\sigma_v^2}-1}$$
$$\times \exp\left\{-\left(\frac{\Theta-\Omega}{\sigma_v^2} + \frac{2\Omega z}{\sigma_v^2}\right) w\right\} dw, \tag{4.74}$$

and hence $f_L(\xi)$ can be readily found by expressing $f_L(\xi)$ as a function of $g(\zeta)$ using (4.72).

Having found $f_L(\xi)$, all that remains is to substitute for $f_L(\xi)$ in (4.68), which requires us to consider the following expressions. Firstly we have

$$J_1 = \exp\left\{\left[\frac{(\alpha - \sigma_v^2)(\Theta - \Omega)}{\sigma_v^2} + \Gamma - i\phi\Psi\right]\tau\right\}$$
$$\times \tilde{U}\left(\phi, \frac{\Theta-\Omega}{\sigma_v^2} + \frac{(\sigma_v^2 s - \Theta + \Omega)2\Omega}{\sigma_v^2[(1-e^{-\Omega\tau})(\sigma_v^2 s - \Theta + \Omega) - 2\Omega e^{-\Omega\tau}]}, 0\right)$$

$$\times\left(\frac{2\Omega e^{-\Omega\tau}}{(\sigma_v^2 s-\Theta+\Omega)(1-e^{-\Omega\tau})+2\Omega e^{-\Omega\tau}}\right)^{\frac{2}{\sigma_v^2}(\sigma_v^2-\alpha)}$$

$$=\exp\left\{\left[\frac{(\alpha-\sigma_v^2)(\Theta-\Omega)}{\sigma_v^2}+\Theta-i\phi\Psi\right]\tau\right\}$$

$$\times\tilde{U}\left(\phi,\frac{\Theta-\Omega}{\sigma_v^2}+\frac{2\Omega(\sigma_v^2 s-\Theta+\Omega)z}{\sigma_v^2[(\sigma_v^2 s-\Theta+\Omega)+2\Omega(z-1)]},0\right)$$

$$\times\left(\frac{2\Omega(z-1)}{(\sigma_v^2 s-\Theta+\Omega)+2\Omega(z-1)}\right)^{2-\frac{2\alpha}{\sigma_v^2}}.$$

Next we consider

$$J_2=\int_z^\infty f_L(\xi)\exp\left\{\left[\frac{(\alpha-\sigma_v^2)(\Theta-\Omega)}{\sigma_v^2}+\Theta-i\phi\Psi\right](\tau-\xi)\right\}\frac{1}{\Omega\zeta^2 e^{-\Omega\xi}}$$

$$\times\left(\frac{2\Omega(z-1)\zeta}{(\sigma_v^2 s-\Theta+\Omega)(\zeta-2)+2\Omega(z-1)\zeta}\right)^{2-\frac{2\alpha}{\sigma_v^2}}d\zeta.$$

By use of (4.72) J_2 becomes

$$J_2=\frac{1}{\Omega}\exp\left\{\left[\frac{(\alpha-\sigma_v^2)(\Theta-\Omega)}{\sigma_v^2}+\Theta-i\Psi\phi\right]\tau\right\}$$

$$\times\int_z^\infty g(\zeta)\left(\frac{2\Omega(z-1)}{(\sigma_v^2 s-\Theta+\Omega)(\zeta-z)+2\Omega(z-1)\zeta}\right)^{2-\frac{2\alpha}{\sigma_v^2}}d\zeta,$$

and substituting for $g(\zeta)$ using (4.74) we have

$$J_2=\frac{1}{\Omega}\exp\left\{\left[\frac{(\alpha-\sigma_v^2)(\Theta-\Omega)}{\sigma_v^2}+\Theta-i\phi\Psi\right]\tau\right\}$$

$$\times\int_z^\infty\frac{-\Omega}{\Gamma\left(\frac{2\alpha}{\sigma_v^2}-1\right)}\int_0^\infty\hat{U}(\phi,w,0)\left(\frac{2\Omega w}{\sigma_v^2}\right)^{\frac{2\alpha}{\sigma_v^2}-1}$$

$$\times\exp\left\{-\left(\frac{\Theta-\Omega}{\sigma_v^2}+\frac{2\Omega\zeta}{\sigma_v^2}\right)w\right\}dw$$

$$\times\left(\frac{2\Omega(z-1)}{(\sigma_v^2 s-\Theta+\Omega)(\zeta-z)+2\Omega(z-1)\zeta}\right)^{2-\frac{2\alpha}{\sigma_v^2}}d\zeta$$

$$=\frac{-[2\Omega(z-1)]^{2-\frac{2\alpha}{\sigma_v^2}}}{\Gamma\left(\frac{2\alpha}{\sigma_v^2}-1\right)}\exp\left\{\left[\frac{(\alpha-\sigma_v^2)(\Theta-\Omega)}{\sigma_v^2}+\Theta-i\phi\Psi\right]\tau\right\}$$

$$\times\int_0^\infty\hat{U}(\phi,w,0)\left(\frac{2\Omega w}{\sigma_v^2}\right)^{\frac{2\alpha}{\sigma_v^2}-1}\exp\left\{-\left(\frac{\Theta-\Omega}{\sigma_v^2}\right)w\right\}J_3(w)dw,\tag{4.75}$$

where for convenience we set

$$J_3(w) = \int_z^\infty e^{-\frac{2\Omega w}{\sigma_v^2}\zeta}[(\sigma_v^2 s - \Theta + \Omega)(\zeta - z) + 2\Omega(z-1)\zeta]^{\frac{2\alpha}{\sigma_v^2}-2} d\zeta.$$

Before proceeding further, we perform some extensive manipulations on $J_3(w)$. Firstly, make the change of integration variable $y = (\sigma_v^2 s - \Theta\Omega)(\zeta - z) + 2\Omega(z-1)\zeta$ to give

$$J_3(w) = \int_{2\Omega(z-1)z}^\infty \exp\left\{\frac{-2\Omega w}{\sigma_v^2}\left(\frac{y + (\sigma_v^2 s - \Theta + \Omega)z}{(\sigma_v^2 s - \Theta + \Omega) + 2\Omega(z-1)}\right)\right\} y^{\frac{2\alpha}{\sigma_v^2}-2}$$

$$\times \frac{dy}{(\sigma_v^2 s - \Theta + \Omega) + 2\Omega(z-1)}$$

$$= \frac{1}{(\sigma_v^2 s - \Theta + \Omega) + 2\Omega(z-1)} \exp\left\{\frac{-2\Omega w(\sigma_v^2 s - \Theta + \Omega)z}{\sigma_v^2[(\sigma_v^2 s - \Theta + \Omega) + 2\Omega(z-1)]}\right\}$$

$$\times \int_{2\Omega(z-1)z}^\infty \exp\left\{\frac{-2\Omega w y}{\sigma_v^2\left[(\sigma_v^2 s - \Theta + \Omega) + 2\Omega(z-1)\right]}\right\} y^{\frac{2\alpha}{\sigma_v^2}-2} dy.$$

By making a further change of integration variable, namely

$$\xi = \frac{2\Omega w y}{\sigma_v^2\left[(\sigma_v^2 s - \Gamma + \Omega) + 2\Omega(z-1)\right]},$$

we have

$$J_3(w) = \frac{\sigma_v^2}{2\Omega w}\left(\frac{\sigma_v^2[(\sigma_v^2 s - \Theta + \Omega) + 2\Omega(z-1)]}{2\Omega w}\right)^{\frac{2\alpha}{\sigma_v^2}-2}$$

$$\times \exp\left\{\frac{-2\Omega w(\sigma_v^2 s - \Theta + \Omega)z}{\sigma_v^2\left[(\sigma_v^2 s - \Theta + \Omega) + 2\Omega(z-1)\right]}\right\}$$

$$\times \int_{\frac{4\Omega^2(z-1)zw}{\sigma_v^2[(\sigma_v^2 s - \Theta + \Omega) + 2\Omega(z-1)]}}^\infty e^{-\xi}\xi^{\left(\frac{2\alpha}{\sigma_v^2}-1\right)-1} d\xi,$$

which in terms of functions (4.21) and (4.22) may be written

$$J_3(w) = [(\sigma_v^2 s - \Theta + \Omega) + 2\Omega(z-1)]^{\frac{2\alpha}{\sigma_v^2}-2}$$

$$\times \exp\left\{\frac{-2\Omega w(\sigma_v^2 s - \Theta + \Omega)z}{\sigma_v^2[(\sigma_v^2 s - t + \Omega) + 2\Omega(z-1)]}\right\}\left(\frac{\sigma_v^2}{2\Omega w}\right)^{\frac{2\alpha}{\sigma_v^2}-1}$$

$$\times \left[\Gamma\left(\frac{2\alpha}{\sigma_v^2}-1\right) - \int_0^{\frac{4\Omega^2(z-1)zw}{\sigma_v^2[(\sigma_v^2 s - \Theta + \Omega) + 2\Omega(z-1)]}} e^{-\xi}\xi^{\left(\frac{2\alpha}{\sigma_v^2}-1\right)-1} d\xi\right] \tag{4.76}$$

Substituting (4.76) into (4.75) we find that

$$J_2 = \frac{-1}{\Gamma\left(\frac{2\alpha}{\sigma_v^2} - 1\right)} \exp\left\{\left[\frac{(\alpha - \sigma_v^2)(\Theta - \Omega)}{\sigma_v^2} + \Theta - i\phi\Psi\right]\tau\right\}$$
$$\times \left(\frac{2\Omega(z-1)}{(\sigma_v^2 s - \Theta + \Omega) + 2\Omega(z-1)}\right)^{2-\frac{2\alpha}{\sigma_v^2}}$$
$$\times \int_0^\infty \hat{U}(\phi, w, 0) e^{-\left(\frac{\Theta - \Omega}{\sigma_v^2}\right)w} \exp\left\{\frac{-2\Omega w(\sigma_v^2 s - \Theta + \Omega)z}{\sigma_v^2[(\sigma_v^2 s - \Theta + \Omega) + 2\Omega(z-1)]}\right\}$$
$$\times \Gamma\left(\frac{2\alpha}{\sigma_v^2} - 1\right)\left[1 - \Gamma\left(\frac{2\alpha}{\sigma_v^2} - 1; \frac{4\Omega^2(z-1)zw}{\sigma_v^2[(\sigma_v^2 s - \Theta + \Omega) + 2\Omega(z-1)]}\right)\right] dw,$$

and since $\tilde{U}(\phi, s, \tau) = J_1 + J_2$, we have

$$\tilde{U}(\phi, s, \tau) = \exp\left\{\left[\frac{(\alpha - \sigma_v^2)(\Theta - \Omega)}{\sigma_v^2} + \Theta - i\phi\Psi\right]\tau\right\}$$
$$\times \left(\frac{2\Omega(z-1)}{(\sigma_v^2 s - \Theta + \Omega) + 2\Omega(z-1)}\right)^{2-\frac{2\alpha}{\sigma_v^2}}$$
$$\times \int_0^\infty \hat{U}(\phi, w, 0) e^{-\left(\frac{\Theta - \Omega}{\sigma_v^2}\right)w}$$
$$\times \exp\left\{\frac{-2\Omega w(\sigma_v^2 s - \Theta + \Omega)z}{\sigma_v^2[(\sigma_v^2 s - \Theta + \Omega) + 2\Omega(z-1)]}\right\}$$
$$\times \Gamma\left(\frac{2\alpha}{\sigma_v^2} - 1; \frac{4\Omega^2(z-1)zw}{\sigma_v^2[(\sigma_v^2 s - \Theta + \Omega) + 2\Omega(z-1)]}\right) dw,$$

which, after substituting for z from (4.70) becomes

$$\tilde{U}(\phi, s, \tau) = \exp\left\{\left[\frac{(\alpha - \sigma_v^2)(\Theta - \Omega)}{\sigma_v^2} + \Theta - i\phi\Psi\right]\tau\right\}$$
$$\times \left(\frac{2\Omega}{(\sigma_v^2 s - \Theta + \Omega)(e^{\Omega\tau} - 1) + 2\Omega}\right)^{2-\frac{2\alpha}{\sigma_v^2}}$$
$$\times \int_0^\infty \hat{U}(\phi, w, 0) e^{-\left(\frac{\Theta - \Omega}{\sigma_v^2}\right)w}$$
$$\times \exp\left\{\frac{-2\Omega w(\sigma_v^2 s - \Theta + \Omega)e^{\Omega\tau}}{\sigma_v^2[(\sigma_v^2 s - \Theta + \Omega)(e^{\Omega\tau} - 1) + 2\Omega]}\right\}$$

$$\times \Gamma\left(\frac{2\alpha}{\sigma_v^2} - 1; \frac{2\Omega w e^{\Omega\tau}}{\sigma_v^2(e^{\Omega\tau} - 1)}\right.$$
$$\left.\times \frac{2\Omega}{(\sigma_v^2 s - \Theta + \Omega)(e^{\Omega\tau} - 1) + 2\Omega}\right) dw,$$

which is the result in Proposition 4.5, after identifying $\tilde{U}(\phi, s, 0)$ with $\tilde{u}(\phi, s)$.

4.F Proof of Proposition 4.6 — Inverting the Laplace Transform

The inverse Laplace transform of (4.19) is most easily found by using the new variables

$$A = \frac{2\Omega w}{\sigma_v^2(1 - e^{-\Omega\tau})}, \quad z = \frac{1}{2\Omega}\{(\sigma_v^2 s - \Theta + \Omega)(e^{\Omega\tau} - 1) + 2\Omega\}. \tag{4.77}$$

If we set

$$h(\phi, w, \tau) = \exp\left\{\left[\frac{(\alpha - \sigma_v^2)(\Theta - \Omega)}{\sigma_v^2} + \Theta - i\phi\Psi\right]\tau\right\} \hat{u}(\phi, w) e^{-\left(\frac{\Theta-\Omega}{\sigma_v{}^2}\right)w}, \tag{4.78}$$

then under the change of variables (4.77) and making use of equation (4.21), equation (4.19) becomes

$$\tilde{U}(\phi, s(z), \tau) = \int_0^\infty h\left(\phi, \frac{\sigma_v^2(1 - e^{-\Omega\tau})A}{2\Omega}, \tau\right)$$
$$\times \exp\left\{\frac{-(\sigma_v^2 s - \Theta + \Omega)(e^{\Omega\tau} - 1)A}{2\Omega z}\right\}$$
$$\times \frac{z^{\frac{2\alpha}{\sigma_v^2} - z}}{\Gamma\left(\frac{2\alpha}{\sigma_v^2} - 1\right)}\left[\frac{\sigma_v^2(1 - e^{-\Omega\tau})}{2\Omega}\right]\int_0^{\frac{A}{z}} e^{-\beta}\beta^{\frac{2\alpha}{\sigma_v^2} - 2} d\beta dA.$$

Changing the integration variable according to $\xi = 1 - \frac{z}{A}\beta$, we have

$$\tilde{U}(\phi, s(z), \tau) = \int_0^\infty h\left(\phi, \frac{\sigma_v^2(1 - e^{-\Omega\tau})A}{2\Omega}, \tau\right)\frac{\sigma_v^2(1 - e^{-\Omega\tau})}{2\Omega}$$
$$\times e^{-A}\frac{A^{\frac{2\alpha}{\sigma_v^2} - 1}}{\Gamma\left(\frac{2\alpha}{\sigma_v^2} - 1\right)}\int_0^1 (1 - \xi)^{\frac{2\alpha}{\sigma_v^2} - 2} z^{-1} e^{\frac{A\xi}{z}} d\xi dA, \tag{4.79}$$

where we make use of the fact that

$$\frac{(\sigma_v^2 s - \Theta + \Omega)(e^{\Omega\tau} - 1)}{2\Omega z} - \frac{A}{z} = \frac{(\sigma_v^2 s - \Theta + \Omega)(e^{\Omega\tau} - 1)A + 2\Omega A}{(\sigma_v^2 s - \Theta + \Omega)(e^{\Omega\tau} - 1) + 2\Omega} = A.$$

In equation (4.16), the Laplace transform is defined with respect to the parameter s. In order to invert (4.79), we must first establish the relationship between the Laplace transform with respect to the parameter s, and the inverse Laplace transform with respect to the parameter z, which is a function of s as defined in the second part of (4.77).

From (4.77) we see that

$$s = \frac{2\Omega(z-1)}{\sigma_v^2(e^{\Omega\tau} - 1)} + \frac{\Theta - \Omega}{\sigma_v^2}.$$

Substituting this into (4.16) gives

$$\mathcal{L}_v\{\hat{U}(\phi, v, \tau)\} = \int_0^\infty \exp\left\{-\left[\frac{2\Omega(z-1)}{\sigma_v^2(e^{\Omega\tau} - 1)} + \frac{\Theta - \Omega}{\sigma_v^2}\right]v\right\}\hat{U}(\phi, v, \tau)dv.$$

By letting

$$y = \frac{2\Omega v}{\sigma_v^2(e^{\Omega\tau} - 1)}, \tag{4.80}$$

we have

$$\begin{aligned}
&\mathcal{L}_v\{\hat{U}(\phi, v(y), \tau)\} \\
&\quad = \frac{\sigma_v^2(e^{\Omega\tau} - 1)}{2\Omega}\int_0^\infty e^{-zy}\exp\left\{-\left(\frac{(\Theta - \Omega)(e^{\Omega\tau} - 1)}{2\Omega} - 1\right)y\right\} \\
&\quad \times \hat{U}(\phi, v(y), \tau)dy,
\end{aligned}$$

which, by use of (4.16) can be written as

$$\begin{aligned}
\mathcal{L}_v\{\hat{U}(\phi, v(y), \tau)\} &= \frac{\sigma_v^2(e^{\Omega\tau} - 1)}{2\Omega}\mathcal{L}_y \\
&\quad \times \left\{\exp\left\{-\left(\frac{(\Theta - \Omega)(e^{\Omega\tau} - 1)}{2\Omega} - 1\right)y\right\}\hat{U}(\phi, v(y), \tau)\right\}.
\end{aligned}$$

Thus we find that

$$\begin{aligned}
\mathcal{L}_v^{-1}\{\tilde{U}(\phi, s(z), \tau)\} &= \frac{2\Omega}{\sigma_v^2(e^{\Omega\tau} - 1)}\exp\left\{\left[\frac{(\Theta - \Omega)(e^{\Omega\tau} - 1)}{2\Omega} - 1\right]y\right\} \\
&\quad \times \mathcal{L}_y^{-1}\{\tilde{U}(\phi, s(z), \tau)\}, \qquad (4.81)
\end{aligned}$$

where

$$\mathcal{L}_y\{f(y)\} = \int_0^\infty e^{-zy} f(y)dy, \tag{4.82}$$

and we recall that y is given by (4.80), and z is defined by (4.77).

Applying the inverse transform (4.81) to (4.79), we have

$$\begin{aligned}\hat{U}(\phi, v(y), \tau) = &\int_0^\infty h\left(\phi, \frac{\sigma_v^2(1-e^{-\Omega\tau})A}{2\Omega}, \tau\right) \frac{\sigma_v^2(1-e^{-\Omega\tau})}{2\Omega} \\ &\times e^{-A} \frac{A^{\frac{2\alpha}{\sigma_v^2}-1}}{\Gamma\left(\frac{2\alpha}{\sigma_v^2}-1\right)} \frac{2\Omega}{\sigma_v^2(e^{\Omega\tau}-1)} \\ &\times \exp\left\{\left[\frac{(\Theta-\Omega)(e^{\Omega\tau}-1)}{2\Omega} - 1\right] y\right\} \\ &\times \int_0^1 (1-\xi)^{\frac{2\alpha}{\sigma_v^2}-2} \mathcal{L}_y^{-1}\left\{z^{-1} e^{\frac{A\xi}{z}}\right\} d\xi dA.\end{aligned}$$

Referring to Abramowitz and Stegun (1970) we find that

$$\mathcal{L}_y^{-1}\{I_0(2\sqrt{A\xi y})\} = \frac{1}{z} e^{\frac{A\xi}{z}},$$

where $I_k(x)$ is the modified Bessel function defined by (4.24). Thus the inverse Laplace transform of $\tilde{U}(\phi, s, \tau)$ becomes

$$\begin{aligned}\hat{U}(\phi, v(y), \tau) = &\int_0^\infty h\left(\phi, \frac{\sigma_v^2(1-e^{-\Omega\tau})A}{2\Omega}, \tau\right) \frac{\sigma_v^2(1-e^{-\Omega\tau})}{2\Omega} \\ &\times e^{-A} \frac{A^{\frac{2\alpha}{\sigma_v^2}-1}}{\Gamma\left(\frac{2\alpha}{\sigma_v^2}-1\right)} \frac{2\Omega}{\sigma_v^2(e^{\Omega\tau}-1)} \\ &\times \exp\left\{\left[\frac{(\Theta-\Omega)(e^{\Omega\tau}-1)}{2\Omega} - 1\right] y\right\} \\ &\times \int_0^1 (1-\xi)^{\frac{2\alpha}{\sigma_v^2}-2} I_0(2\sqrt{A\xi y}) d\xi dA.\end{aligned}$$

We can further simplify this result by noting that[8]

$$\int_0^1 (1-\xi)^{\frac{2\alpha}{\sigma_v^2}-2} I_0(2\sqrt{A\xi y}) d\xi = \Gamma\left(\frac{2\alpha}{\sigma_v^2}-1\right) (Ay)^{\frac{1}{2}-\frac{\alpha}{\sigma_v^2}} I_{\frac{2\alpha}{\sigma_v^2}-1}(2\sqrt{Ay}),$$

[8]This result is simply obtained by expanding both terms in the integral in power series.

and therefore

$$
\begin{aligned}
&\hat{U}(\phi, v, \tau) \\
&= \int_0^\infty h\left(\phi, \frac{\sigma_v^2(1-e^{-\Omega\tau})A}{2\Omega}, \tau\right) \frac{\sigma_v^2(1-e^{-\Omega\tau})}{2\Omega} \frac{2\Omega}{\sigma_v^2(e^{\Omega\tau}-1)} \\
&\quad \times e^{-A-y}\left(\frac{A}{y}\right)^{\frac{\alpha}{\sigma_v^2}-\frac{1}{2}} \exp\left\{\frac{(\Theta-\Omega)(e^{\Omega\tau}-1)}{2\Omega}y\right\} I_{\frac{2\alpha}{\sigma_v^2}-1}(2\sqrt{Ay})dA.
\end{aligned}
$$

Recalling the definitions for A and y, from (4.77) and (4.80) respectively, we conclude that

$$
\begin{aligned}
&\hat{U}(\phi, v, \tau) \\
&= \int_0^\infty h(\phi, w, \tau) \frac{2\Omega}{\sigma_v^2(e^{\Omega\tau}-1)} \exp\left\{-\frac{2\Omega}{\sigma_v^2(e^{\Omega\tau}-1)}(we^{\Omega\tau}+v)\right\} \\
&\quad \times \left(\frac{we^{\Omega\tau}}{v}\right)^{\frac{\alpha}{\sigma_v^2}-\frac{1}{2}} \exp\left\{\frac{(\Theta-\Omega)}{\sigma_v^2}v\right\} I_{\frac{2\alpha}{\sigma_v^2}-1} \\
&\quad \times \left(\frac{4\Omega}{\sigma_v^2(e^{\Omega\tau}-1)}(wve^{\Omega\tau})^{\frac{1}{2}}\right) dw.
\end{aligned}
$$

Finally, substituting for $h(\phi, w, \tau)$ from (4.78) we obtain the result in Proposition 4.6.

4.G Proof of Proposition 4.7 — Inverting the Fourier Transform

Taking the inverse Fourier transform of $\hat{U}(\phi, v, \tau)$ in (4.23) and returning to the original variables, we find that

$$
C_E(S, v, \tau) = e^{-r\tau} \int_0^\infty \int_{-\infty}^\infty c(e^y, w)\bar{G}(y, w, \tau; S, v)dydw,
$$

where we identify $\bar{G}(y, w, \tau; S, v)$ as

$$
\begin{aligned}
&\bar{G}(y, w, \tau; S, v) \\
&= \frac{1}{2\pi}\int_{-\infty}^\infty e^{i\phi y} e^{\frac{(\Theta-\Omega)}{\sigma_v^2}(v-w+\alpha\tau)} e^{-i\phi(r-q-\lambda^* k^*)\tau} e^{\gamma[A(\phi)-1]\tau} e^{-i\phi\ln S} \\
&\quad \times \frac{2\Omega}{\sigma_v^2(e^{\Omega\tau}-1)}\left(\frac{we^{\Omega\tau}}{v}\right)^{\frac{\alpha}{\sigma_v^2}-\frac{1}{2}} \exp\left\{\frac{-2\Omega}{\sigma_v^2(e^{\Omega\tau}-1)}(we^{\Omega\tau}+v)\right\} \\
&\quad \times I_{\frac{2\alpha}{\sigma_v^2}-1}\left(\frac{4\Omega}{\sigma_v^2(e^{\Omega\tau}-1)}(wve^{\Omega\tau})^{\frac{1}{2}}\right) d\phi.
\end{aligned}
$$

Next we expand the term $e^{\lambda^* A(\phi)\tau}$ using a Taylor series expansion, and find that

$$
\begin{aligned}
e^{\lambda^* A(\phi)\tau} &= \sum_{n=0}^{\infty} \frac{[\lambda^* \tau]^n}{n!}[A(\phi)]^n \\
&= \sum_{n=0}^{\infty} \frac{[\gamma\tau]^n}{n!} \int_0^\infty \int_0^\infty \cdots \int_0^\infty e^{-i\phi \ln Y_1 Y_2 \ldots Y_n} \\
&\quad \times G^*(Y_1)G^*(Y_2)\ldots G^*(Y_n)dY_1 dY_2 \ldots dY_n,
\end{aligned}
$$

where each $Y_j, j = 0, \ldots, n$ is an independent jump drawn from the density $G^*(Y)$.

We define $X_n \equiv Y_1 Y_2, Y_n$, and $X_0 \equiv 1$, and assume that $G^*(Y)$ is of a form that allows us to make a simplification of the form

$$
\begin{aligned}
&\int_0^\infty \int_0^\infty \cdots \int_0^\infty f(Y_1 Y_2 \ldots Y_n) G^*(Y_1)G^*(Y_2)\ldots G^*(Y_n)dY_1 dY_2 \ldots dY_n \\
&= \int_0^\infty f(X_n)G^*(X_n)dX_n \equiv \mathbb{E}_{\mathbb{Q}^*}^{(n)}[F(X_n)],
\end{aligned}
$$

where $f(X_n)$ is some general function of X_n. Thus we have

$$
e^{\lambda^* A(\phi)\tau} = \sum_{n=0}^{\infty} \frac{(\lambda^*\tau)^n}{n!} \mathbb{E}_{\mathbb{Q}^*}^{(n)}[e^{-i\phi \ln X_n}],
$$

and hence the Green's function can be written as it appears in (4.27). Finally, (4.26) is obtained by substituting for $c(e^y, w)$ using (2.19).

4.H Proof of Proposition 4.8 — Deriving the Price for a European Call

In order to derive the price of the European call option, $C_E(S, v, \tau)$, written on S, we begin by considering the integral

$$
\begin{aligned}
&\int_0^\infty \bar{G}(y, w, \tau; S, v)dw \\
&= \sum_{n=0}^{\infty} \frac{(\lambda^*\tau)^n e^{-\lambda^*\tau}}{n!} \mathbb{E}_{\mathbb{Q}^*}^{(n)} \left\{ \frac{1}{2\pi} \int_{-\infty}^{\infty} e^{i\phi y} e^{-i\phi \ln(S X_n e^{-\lambda^* k^* \tau})} \right. \\
&\quad \times \left[\int_0^\infty e^{\frac{(\Theta - \Omega)}{\sigma_v^2}(v - w + \alpha\tau)} e^{-i\phi(r-q)\tau} \frac{2\Omega}{\sigma_v^2(e^{\Omega\tau} - 1)} \left(\frac{w e^{\Omega\tau}}{v} \right)^{\frac{\alpha}{\sigma_v^2} - \frac{1}{2}} \right.
\end{aligned}
$$

$$\times \exp\left\{-\frac{2\Omega}{\sigma_v^2(e^{\Omega\tau}-1)}(we^{\Omega\tau}+v)\right\} I_{\frac{2\alpha}{\sigma_v^2}-1}$$
$$\times \left(\frac{4\Omega}{\sigma_v^2(e^{\Omega\tau}-1)}(wve^{\Omega\tau})^{\frac{1}{2}}\right) dw\Bigg] d\phi\Bigg\}.$$

Evaluating the integral with respect to u, we have

$$\int_0^\infty \bar{G}(y,w,\tau;S,v)dw$$
$$= \sum_{n=0}^{\infty} \frac{(\lambda^*\tau)^n e^{-\lambda^*\tau}}{n!}$$
$$\times \mathbb{E}_{\mathbb{Q}^*}^{(n)}\left\{\frac{1}{2\pi}\int_{-\infty}^{\infty} e^{i\phi y} e^{-i\phi \ln SX_n e^{-\gamma k^*\tau}} e^{B_2(-\phi,\tau)+D_2(-\phi,\tau)v} d\phi\right\},$$

where

$$B_2(\phi,\tau) = i\phi(r-q)\tau + \frac{\alpha}{\sigma_v^2}\left\{(\Theta_2+\Omega_2)\tau - 2\ln\left(\frac{1-Q_2e^{\Omega\tau}}{1-Q_2}\right)\right\},$$

and

$$D_2(\phi,\tau) = \frac{(\Theta_2+\Omega_2)}{\sigma_v^2}\left[\frac{1-e^{\Omega_2\tau}}{1-Q_2e^{\Omega_2\tau}}\right],$$

where we define $\Theta_2 = \Theta_2(\phi) \equiv \Theta(-\phi)$, $\Omega_2 = \Omega_2(\phi) \equiv \Omega(-\phi)$, and $Q_2 = Q_2(\phi) \equiv (\Theta_2+\Omega_2)/(\Theta_2-\Omega_2)$.

Referring to the results given by Adolfsson *et al.* (2013), it follows that

$$C_E(S,v,\tau)$$
$$= e^{-r\tau}\int_{\ln K}^{\infty}(e^y-K)\Bigg[\sum_{n=0}^{\infty}\frac{(\lambda^*\tau)^n e^{-\lambda^*\tau}}{n!}$$
$$\times \mathbb{E}_{\mathbb{Q}^*}^{(n)}\left\{\frac{1}{2\pi}\int_{-\infty}^{\infty} e^{i\phi y} e^{-i\phi \ln SX_n e^{-\lambda^* k^*\tau}} e^{B_2(-\phi,\tau)+D_2(-\phi,\tau)v} d\phi\right\}\Bigg] dy$$
$$= \frac{e^{-r\tau}}{2\pi}\sum_{n=0}^{\infty}\frac{(\lambda^*\tau)^n e^{-\lambda^*\tau}}{n!}$$
$$\times \mathbb{E}_{\mathbb{Q}^*}^{(n)}\left\{\int_{-\infty}^{\infty} f_2(SX_ne^{-\lambda^*k^*\tau},v,\tau;-\phi)\int_{\ln K}^{\infty}(e^y-K)e^{i\phi y}dyd\phi\right\},$$

where we set

$$f_2(SX_ne^{-\lambda^*k^*\tau},v,\tau;\phi) \equiv e^{B_2(\phi,\tau)+D_2(\phi\tau)v+i\phi\ln SX_ne^{-\lambda^*k^*\tau}}.$$

Further use of the results from Adolfsson *et al.* (2013) allows us to express $C_E(S, v, \tau)$ as

$$C_E(S, v, \tau) = \sum_{n=0}^{\infty} \frac{(\lambda^*\tau)^n e^{-\lambda^*\tau}}{n!} \mathbb{E}_{\mathbb{Q}^*}^{(n)}$$
$$\times\{SX_n e^{-\lambda^* k^* \tau} e^{-q\tau} P_1^H(SX_n e^{-\lambda^* k^* \tau}, v, \tau; K)$$
$$-Ke^{-r\tau} P_2^H(SX_n e^{-\lambda^* k^* \tau}, v, \tau; K)\},$$

where $P_j^H(SX_n e^{-\lambda^* k^* \tau}, v, \tau; K)$, for $j = 1, 2$, is as defined in (4.30).

4.I Deriving the Early Exercise Premium

From Duhamel's principle in Proposition 4.2 we know that

$$C_P(S, v, \tau)$$
$$= \int_0^\tau e^{-r(\tau-\xi)} \int_0^\infty \int_{\ln a(w,\xi)}^\infty f(e^y, w, \xi)\bar{G}(y, w, \tau - \xi; S, v)dydud\xi,$$

where $f(e^y, w, \xi)$ is given by (4.6), and $\bar{G}(y, w, \tau - \xi; S, v)$ is given by (4.27). To simplify $C_P(S, v, \tau)$ we consider two linear components: $C_P^D(S, v, \tau)$, which is the early exercise premium due to diffusion in S, and $C_P^J(S, v, \tau)$, the cost incurred by the option holder due to downward jumps. Thus we have

$$C_P(S, v, \tau) = C_P^D(S, v, \tau) - \lambda^* C_P^J(S, v, \tau),$$

where

$$C_P^D(S, v, \tau)$$
$$= \int_0^\tau e^{-r(\tau-\xi)} \int_0^\infty \int_{\ln a(w,\xi)}^\infty (qe^y - rK)\bar{G}(y, w, \tau - \xi; S, v)dydud\xi, \tag{4.83}$$

and

$$C_P^J(S, v, \tau) = \int_0^\tau e^{-r(\tau-\xi)} \int_0^\infty \int_{\ln a(w,\xi)}^\infty \int_0^{a(w,\xi)e^{-y}}$$
$$\times [C^A(e^y Y, w, \tau) - (e^y Y - K)G(Y)]dY$$
$$\times \bar{G}(y, w, \tau - \xi; S, v)dydud\xi. \tag{4.84}$$

Firstly we consider $C_P^D(S, v, \tau)$. Substituting for $\bar{G}$ we have

$$C_P^D(S, v, \tau) = \sum_{n=0}^{\infty} \frac{(\lambda^*\tau)^n e^{-\lambda^*\tau}}{n!}$$

$$\times \mathbb{E}_{\mathbb{Q}^*}^{(n)}\left\{ \int_0^\tau (\tau-\xi)^n e^{-\lambda^*(\tau-\xi)} e^{-r(\tau-\xi)} \int_0^\infty \int_{\ln a(w,\xi)}^\infty [qe^y - rK] \right.$$

$$\times \frac{1}{2\pi} \int_{-\infty}^\infty e^{i\phi y} e^{\frac{(\Theta-\Omega)}{\sigma_v^2}(v-w+\alpha(\tau-\xi))} e^{-i\phi(r-q)(\tau-\xi)}$$

$$\times e^{-i\phi \ln SX_n e^{-\lambda^* k^*(\tau-\xi)}} \frac{2\Omega}{\sigma_v^2(e^{\Omega(\tau-\xi)}-1)} \left(\frac{we^{\Omega(\tau-\xi)}}{v}\right)^{\frac{\alpha}{\sigma_v^2}-\frac{1}{2}}$$

$$\times \exp\left\{ -\frac{2\Omega}{\sigma_v^2(e^{\Omega(\tau-\xi)}-1)}(we^{\Omega(\tau-\xi)}+v)\right\}$$

$$\left. \times I_{\frac{2\alpha}{\sigma_v^2}-1}\left(\frac{4\Omega}{\sigma_v^2(e^{\Omega(\tau-\xi)}-1)}(wve^{\Omega(\tau-\xi)})^{\frac{1}{2}}\right) d\phi dy dw d\xi \right\}.$$

Using results from Adolfsson *et al.* (2013), we can readily show that

$$C_P^D(S,v,\tau)$$
$$= \sum_{n=0}^\infty \frac{(\lambda^*\tau)^n e^{-\lambda^*\tau}}{n!} \mathbb{E}_{\mathbb{Q}^*}^{(n)}\left\{ \int_0^\tau \int_0^\infty (\tau-\xi)^n e^{-\lambda^*(\tau-\xi)} \right.$$
$$\times [qSX_n e^{-\lambda^* k^*(\tau-\xi)} e^{-q(\tau-\xi)} P_1^A(SX_n e^{-\lambda^* k^*(\tau-\xi)}, v, \tau-\xi, w, a(w,\xi))$$
$$\left. - rKe^{-r(\tau-\xi)} P_2^A(SX_n e^{-\lambda^* k^*(\tau-\xi)}, v, \tau-\xi, w, a(w,\xi))] dw dy \right\},$$

where $P_j^A(SX_n e^{-\lambda^* k^*(\tau-\xi)}, v, \tau-\xi, w, a(w,\xi))$ is given by (4.34).

Next we examine the $C_P^J(S,v,\tau)$ term. Interchanging the order of integration with respect to Y and y, we have

$$C_P^J(S,v,\tau) = \int_0^\tau e^{-r(\tau-\xi)} \int_0^\infty \int_0^1 G^*(Y) \int_{\ln a(w,\xi)}^{\ln \frac{a(w,\xi)}{Y}}$$
$$\times [C^A(e^y Y, w, \tau) - (e^y Y - K)]$$
$$\times \bar{G}(y, w, \tau-\xi; S, v) dy dY dw d\xi.$$

Making the change of integration variable $z = e^y$, we have

$$C_P^J(S,v,\tau)$$
$$= \int_0^\tau \int_0^\infty e^{-r(\tau-\xi)} \int_0^1 G^*(Y) \int_{a(w,\xi)}^{\frac{a(w,\xi)}{Y}} \frac{[C^A(zY,w,\xi) - (zY-K)]}{z}$$
$$\times \bar{G}(\ln z, w, \tau-\xi; S, v) dz dY dw d\xi$$

$$= \sum_{n=0}^{\infty} \frac{(\lambda^* \tau)^n e^{-\lambda^* \tau}}{n!} \mathbb{E}_{\mathbb{Q}^*}^{(n)} \left\{ \int_0^\tau \int_0^\infty (\tau - \xi)^n e^{-\lambda^* k^* (\tau - \xi)} e^{-r(\tau - \xi)} \right.$$

$$\times \int_0^1 G(Y) \int_{a(w,\xi)}^{a(w,\xi)/Y} [C^A(zY, w, \xi) - (zY - K)]$$

$$\left. \times \bar{Q}_J(z, w, \tau - \xi; SX_n e^{-\lambda^* k^* (\tau - \xi)}, v) dz dY dw d\xi \right\},$$

where $\bar{Q}_J(z, w, \tau - \xi; SX_n e^{-\lambda^* k^* (\tau - \xi)}, v)$ is defined in (4.37). This is seen to be equivalent to the statement in Proposition 4.9.

Chapter 5

Representation and Numerical Approximation of American Option Prices under Heston

5.1 Introduction

In this chapter we seek to generalise the constant volatility analysis of American option pricing to the stochastic volatility case, focusing in particular on Heston's square root model using transform methods. We derive the solution by first posing the problem in the form proposed by Jamshidian (1992) for American option pricing. As we have explained in the previous chapter in the case of stochastic volatility and jumps, this involves converting the American option pricing problem from one involving the solution of a homogeneous partial differential equation (PDE) (since there is no jump term in this chapter the partial-integro differential equation becomes simply a partial differential equation) on a finite domain to one involving the solution to an inhomogeneous partial differential equation (PDE) in an infinite domain. As in the previous chapter, and making adjustments to cater for the fact that we do not have a jump term, we represent the solution of the inhomogeneous PDE as the sum of solution to the homogeneous part and the corresponding inhomogeneous term by taking advantage of the Duhamel's principle. We then solve the PDE associated with the density function using Fourier and Laplace transforms. In this way we incidentally demonstrate how Heston's results for the European call, derived by him using characteristic functions, may also be found by means of Fourier and Laplace transforms, without the need to assume the form of the solution. We numerically solve the resulting integral equation system for the free surface using an iterative method, based on an approximation of the early exercise surface proposed by Tzavalis and Wang (2003).

Stochastic volatility models for pricing derivative securities have been developed as an extension to the original, constant volatility model of Black and Scholes (1973). Many studies of market prices for option contracts have consistently found that implied volatilities vary with respect to both maturity and the moneyness of the option. Directly modelling the volatility of the underlying asset using an additional stochastic process provides one means by which this empirical feature can be incorporated into the Black-Scholes pricing framework.

There are a number of methods that one can use to model volatility stochastically. Hull and White (1987) model the variance using a geometric Brownian motion, as well as an Ornstein-Uhlenbeck process with mean-reversion related to the volatility. In practice, mean-reversion is considered to be an essential feature of observed volatility, and thus most plausible models are of the Ornstein-Uhlenbeck type. Wiggins (1987) models the logarithm of the volatility with mean-reversion, whereas Scott (1987), Johnson and Shanno (1987), Heston (1993) and Stein and Stein (1991) model the variance using a square root process. Zhu (2000) also considers a double square root process, which is an extension of the basic square root process in which both the drift and diffusion coefficients involve the volatility. In this chapter we focus on Heston's square root model, which seems to have become the dominant stochastic volatility model in the literature, one reason perhaps being that Heston (1993) was able to provide a convenient analytic expression for European option prices.

Pricing American options under stochastic volatility has much in common with the standard geometric Brownian motion case. In the constant volatility case, it is well known that the price of an American call option can be decomposed into the sum of a corresponding European call and an early exercise premium term. Kim (1990), Jacka (1991b) and Carr *et al.* (1992) demonstrate this result using a variety of approaches. The American call price takes the form of an integral equation involving the unknown early exercise boundary. By evaluating this equation at the free boundary, a corresponding integral equation for the early exercise condition is produced. As we have seen in the previous chapter (but there also with jumps), this characterization also holds when the volatility is allowed to evolve randomly.

Since the cash flows arising from early exercise are not explicit functions of the volatility, the main complexity arising from generalising American option pricing theory from the constant volatility case to that of stochastic volatility is related to the early exercise boundary. Lewis (2000) indicates that the free boundary becomes a two-dimensional free surface, in which the early exercise value of the underlying is a function of time to maturity and the volatility level. This result is reiterated in a more general context by Detemple and Tian (2002). Further analysis is provided by Touzi (1999), who proves a number of fundamental properties for the free boundary and option price under stochastic volatility for the American put, focusing on how the surface changes with respect to the volatility.

The task of solving the two-dimensional partial differential equation (PDE) for the early exercise premium is obviously more involved than in the one-dimensional case. For European calls, Heston (1993) uses characteristic functions to derive the pricing equation under stochastic volatility. This is closely related to the Fourier transform approach, which was applied to American option pricing, originally by McKean (1965), and used by Chiarella and Ziogas (2005) to consider American strangles. While the Fourier transform, taken with respect to the log of the underlying asset price, is sufficient to solve the problem under constant volatility, it turns out that in addition one needs to take the Laplace transform with respect to the volatility variable. Feller (1951) used the Laplace transform to solve the Fokker-Planck equation for the transition density of the square-root process. Here we use the Laplace transform in conjunction with the Fourier transform to derive the integral equation for the price of an American call under Heston's stochastic volatility model. Subsequently, a corresponding integral equation for the early exercise surface is readily determined.

Since the expression for the early exercise premium for the American call involves a three-dimensional integral, it is generally considered too cumbersome for numerical solution methods. Fouque *et al.* (2000) (see also Fouque *et al.* (2003)) use a series expansion based around the square-root of the rate of mean reversion for the volatility. This method provides an adjustment to the Black-Scholes solution, taking advantage of the fact that the mean reversion feature dominates the volatility dynamics over a sufficiently large time period. The downside to this approach is that the expansion is poor close to expiry and near the early exercise surface, since the mean reversion will not dominate the volatility dynamics in these situations.

The most common numerical solution methods in the literature on American option pricing under stochastic volatility involve a discrete grid or lattice. These are typically used to estimate the distribution of the underlying asset, or to numerically solve the partial differential equation for the option price.

Ikonen and Toivanen (2004) solve the second order PDE for the American put option under stochastic volatility using several implicit finite difference schemes for the space variables. They apply an operator splitting technique for the time component, and the method is easier to compute than fully implicit schemes. Multigrid methods are a related technique which involve solving a problem using a finite difference scheme on successively coarser grids to better control the error, and then interpolating the results back up to the original fine grid. The advantage is that less computation is required to solve the problem for a high level of accuracy, and this is of particular interest for multidimensional problems. Clarke and Parrott (1999), Oosterlee (2003) and Reisinger and Wittum (2004) all provide applications of these methods to option pricing under stochastic volatility. Furthermore, Zvan *et al.* (1998) solve the problem using a hybrid method involving finite elements for the diffusion terms and finite volume for the convection terms.

It was Tzavalis and Wang (2003) who provided one of the few implemented algorithms based on numerical integration methods. In their approach the logarithm of the early exercise boundary is approximated using a Taylor series expansion around the long-run volatility. They cite the empirical findings of Broadie *et al.* (2000) as justification for this assumption, and this leads to a simplified expression for the early exercise premium which is easier to evaluate numerically. To solve the resulting integral equation system, Tzavalis and Wang (2003) approximate the free boundary using Chebyshev polynomials, and are able to produce a fast approximation that leads to accurate option prices. In this chapter we propose a related numerical method, in which we forego approximating the free boundary with respect to time until maturity, and consider instead a different set of fitting volatilities for the free boundary estimate. We use an iterative numerical integration scheme to estimate the free boundary, based on solution techniques typically applied to Volterra integral equations.

The remainder of this chapter is structured as follows. Sec. 5.2 recalls from the previous chapter, in the case of stochastic volatility with jumps, the pricing problem for an American call option under Heston's square root process. We also recall the Jamshidian (1992) formulation for the American call, and the use of Duhamel's principle to determine the form of the solution as the sum of the corresponding European price and the early exercise premium. These expressions are integrals involving the transition density function for the joint stock price and variance processes, so Sec. 5.3 uses integral transform methods to provide a general solution to the PDE for this quantity. We then derive an analytic integral equation for the option price and early exercise boundary in Sec. 5.4, and propose a series expansion approximation for the early exercise boundary in Sec. 5.5. We relate this to the series expansion of Tzavalis and Wang (2003), and outline an iterative scheme for solving the resulting linked system of integral equations for the early exercise surface. Concluding remarks are presented in Sec. 5.6. Most of the lengthy mathematical derivations are given in appendices.

5.2 Problem Statement — The Heston Model

Let $C^A(S, v, \tau)$ be the price of an American call option written on a dividend paying underlying asset of price S with time-to-maturity τ[1] and strike price K. The payoff at maturity of the contract is denoted as $(S - K)^+$. The dynamics of S under the real world measure, $\mathbb{P}$, are driven by the stochastic differential equation (this is equation (2.1) with the jump intensity set to zero),

$$dS = \mu S dt + \sqrt{v} S dZ_1, \tag{5.1}$$

where μ is the instantaneous return per unit time, v is the instantaneous variance (squared volatility) per unit time of the return on the underlying asset and Z_1 is a standard Wiener process. We also allow v to evolve stochastically using the square root process proposed by Heston (1993), so that under the real world measure, the dynamics of v are governed by (this is the same as 2.2)

$$dv = \kappa_v(\theta - v)dt + \sigma_v \sqrt{v} dZ_2, \tag{5.2}$$

where θ is the long-run mean of v, κ_v is the speed of mean-reversion, σ_v is the instantaneous volatility of v per unity time, and Z_2 is a standard Wiener process correlated with Z_1 such that $\mathbb{E}(dZ_1 dZ_2) = \rho dt$.

Now given equations (2.10) and (2.11) we want to come up with the corresponding American call option pricing partial differential equation (PDE). Using the well established hedging techniques and Ito's lemma, we can represent the pricing PDE as,

$$\begin{aligned} \frac{\partial C^A}{\partial \tau} &= (r - q)S\frac{\partial C^A}{\partial S} + \frac{1}{2}vS^2\frac{\partial^2 C^A}{\partial S^2} + [\kappa_v(\theta - v) - \lambda v]\frac{\partial C^A}{\partial v} \\ &\quad + \frac{1}{2}\sigma_v^2 v\frac{\partial^2 C^A}{\partial v^2} + \rho\sigma_v vS\frac{\partial^2 C^A}{\partial S \partial v} - rC^A, \end{aligned} \tag{5.3}$$

where $0 < S < a(v, \tau)$ and $0 < v < \infty$. Here $a(v, \tau)$ is the early exercise surface at variance, v and time-to-maturity, τ. The PDE (5.3) is to be solved

[1] Here, $\tau = T - t$.

subject to the initial and boundary conditions,

$$C^A(S,v,0) = S - K, \quad 0 < S < \infty, \tag{5.4}$$

$$C^A(0,v,\tau) = 0, \tag{5.5}$$

$$C^A(a(v,\tau),v,\tau) = a(v,\tau) - K, \tag{5.6}$$

$$\lim_{S\to a(v,\tau)} \frac{\partial C^A}{\partial S}(S,v,\tau) = 1, \tag{5.7}$$

$$\lim_{S\to a(v,\tau)} \frac{\partial C^A}{\partial v}(S,v,\tau) = 0. \tag{5.8}$$

Condition (5.4) is the payoff at maturity of the option contract while equation (5.5) is the absorbing state which ensures the option ceases to exist once the underlying asset price hits zero. Equation (5.6) is the value matching condition which ensures continuity of the option value function at the early exercise surface, $a(v,\tau)$. Equation (5.7) is called the smooth-pasting condition, which together with the value matching condition are imposed to avoid arbitrage opportunities. Equation (5.8) specifies that the first derivative of the value function with respect to the volatility variable is zero and has been given a number of times before.

As noted in the PDE (5.3), the underlying asset is bounded above by the early exercise boundary, $a(v,\tau)$. Jamshidian (1992) shows how the homogeneous PDE (5.3) can be transformed so that S can have an unbounded domain by introducing an indicator function transforming it to. This is proved in Appendix 4.A in the case of jumps and stochastic volatility. The reader should keep in mind that here the jump term is zero. We give the details just for completeness

$$\begin{aligned}\frac{\partial C^A}{\partial \tau} &= (r-q)S\frac{\partial C^A}{\partial S} + \frac{1}{2}vS^2\frac{\partial^2 C^A}{\partial S^2} + [\kappa_v(\theta - v) - \lambda v]\frac{\partial C^A}{\partial v} \\ &\quad + \frac{1}{2}\sigma_v^2 v\frac{\partial^2 C^A}{\partial v^2} + \rho\sigma_v vS\frac{\partial^2 C^A}{\partial S\partial v} - rC^A + \mathbb{1}_{S\geq a(v,\tau)}(qS - rK),\end{aligned} \tag{5.9}$$

where, $0 < S < \infty$, $0 < v < \infty$, and $\mathbb{1}_{S\geq a(S,v,\tau)}$ is an indicator function.

Also associated with the two stochastic differential equations (2.10) and (2.11) is a bivariate transition density function $G(S,v,\tau;S_0,v_0)$ which denotes the transition from S,v at time-to-maturity τ to S_0,v_0 at maturity. This function is also known as the Green's function and will be used to represent the solution of the PDE (2.12), which is now a PDE because there is no jump term present. The transition density function satisfies the backward Komogorov PDE associated with S and v which is,

$$\frac{\partial G}{\partial \tau} = (r-q)S\frac{\partial G}{\partial S} + \frac{vS^2}{2}\frac{\partial^2 G}{\partial S^2} + [\kappa_v(\theta - v) - \lambda v]\frac{\partial G}{\partial v} + \frac{\sigma_v^2 v}{2}\frac{\partial^2 G}{\partial v^2} + \rho\sigma_v vS\frac{\partial^2 G}{\partial S \partial v}. \tag{5.10}$$

Equation (5.10) is to be solved subject to the initial condition,

$$G(S, v, 0; S_0, v_0) = \delta(S - S_0)\delta(v - v_0),$$

where $\delta(\cdot)$ is the Dirac delta function.

Equation (5.9) is an inhomogeneous PDE the solution of which can be represented by use of Duhamel's principle. For convenience, we let $S = e^x$, $V_A(x, v, \tau) \equiv C^A(e^x, v, \tau)$ and $H(x, v, \tau; u, w) \equiv G(S, v, \tau; S_0, v_0)$ such that the PDE (5.9) becomes,

$$\frac{\partial V_A}{\partial \tau} = \left(r - q - \frac{1}{2}v\right)\frac{\partial V_A}{\partial x} + \frac{1}{2}v\frac{\partial^2 V_A}{\partial x^2} + [\kappa_v(\theta - v) - \lambda v]\frac{\partial V_A}{\partial v} + \frac{\sigma_v^2 v}{2}\frac{\partial^2 V_A}{\partial v^2} + \rho\sigma_v v\frac{\partial^2 V_A}{\partial x \partial v} - rV_A + \mathbb{1}_{x \geq \ln a(v,\tau)}(qe^x - rK). \tag{5.11}$$

Proposition 5.1. *By letting $V_E(x, v, \tau)$ be the solution to the homogeneous part of* (5.11) *and use of Duhamel's principle, it can then be shown that,*

$$V_E(x, v, \tau) = e^{-r\tau}\int_0^\infty \int_{-\infty}^\infty (e^u - K)^+ H(x, v, \tau; u, w)du dw. \tag{5.12}$$

Duhamel's principle states that if the solution to the homogeneous part of the PDE (5.11) *is given by* (5.12), *then the American call option value can be represented as,*

$$V_A(x, v, \tau) = V_E(x, v, \tau) + V_P(x, v, \tau), \tag{5.13}$$

where,

$$V_P(x, v, \tau) = \int_0^\tau e^{-r(\tau-\xi)}\int_0^\infty \int_{\ln a(w,\xi)}^\infty (qe^u - rK)H(x, v, \tau - \xi; u, w)du dw d\xi, \tag{5.14}$$

is the early exercise premium.

Proof. Refer to Appendix 4.B in Chapter 4. The reader should keep in mind that in Chapter 3 we also have the jump term whereas, now we do not have the jump term and so the derivation is simpler. □

By effecting the transformations indicated just before Proposition 5.1, it can be shown that the Kolmogorov PDE becomes,

$$\begin{aligned}\frac{\partial H}{\partial \tau} &= \frac{v}{2}\frac{\partial^2 H}{\partial x^2} + \rho\sigma_v v\frac{\partial^2 H}{\partial x \partial v} + \frac{\sigma_v^2 v}{2}\frac{\partial^2 H}{\partial v^2} \\ &\quad + \left(r - q - \frac{v}{2}\right)\frac{\partial H}{\partial x} + (\alpha - \beta v)\frac{\partial H}{\partial v},\end{aligned} \tag{5.15}$$

where $-\infty < x < \infty$, $0 < v < \infty$, $0 \le \tau \le T$, with $\alpha = \kappa_v\theta$ and $\beta = \kappa_v + \lambda$. The PDE (5.15) is to be solved subject to the initial condition,

$$H(x, v, 0; x_0, v_0) = \delta(e^x - e^{x_0})\delta(v - v_0). \tag{5.16}$$

For convenience, we will just write $H(x, v, 0)$ to mean $H(x, v, 0; u, w)$. Since x does not appear in the coefficients for any terms in (5.15), we can apply Fourier transform to the x-variable and this is presented in Proposition 5.2. It should be noted that integral transforms require knowledge of the behavior of the functions being transformed at the extremities of the domain. In applying the Fourier transform, we require that $H(x, v, \tau)$ and $\partial H/\partial x$ tend to zero as $x \to \pm\infty$. By their nature, density functions tend to zero as the state variables tend to infinity thus it is appropriate to apply transform methods. It is of course also possible to apply the integral transform approach directly to the option pricing equation, but then in order to handle the extremities of the domain we would have to consider the damped option price approach of Carr and Madan (1999), and as we have discussed in earlier chapters.

5.3 Finding the Density Function using Integral Transforms

In this section we derive the explicit form of the density function by solving the PDE (5.15). As highlighted before, this will be accomplished by applying the Fourier transform to the logarithm of the underlying asset variable and the Laplace transform to the stochastic variance variable. Applying these transform techniques transforms the PDE to a corresponding system of ordinary differential equations known as the characteristic equations which are then solved by the method of characteristics. These steps are given by the propositions below.

Proposition 5.2. *Let $\mathcal{F}_x\{H(x,v,\tau)\}$ be the Fourier transform of $H(x,v,\tau)$ taken with respect to x, defined as*

$$\mathcal{F}_x\{H(x,v,\tau)\} = \int_{-\infty}^{\infty} e^{i\phi x} H(x,v,\tau)dx \equiv \hat{H}(\phi,v,\tau). \tag{5.17}$$

Applying the Fourier transform to (5.15) *we find that $\hat{H}$ satisfies the PDE*

$$\frac{\partial \hat{H}}{\partial \tau} = \frac{\sigma_v^2 v}{2}\frac{\partial^2 \hat{H}}{\partial v^2} + (\alpha - \Theta v)\frac{\partial \hat{H}}{\partial v} + \left(\frac{\Lambda}{2}v - i\phi(r-q)\right)\hat{H}, \tag{5.18}$$

with initial condition $\hat{H}(\phi,v,0) = \hat{u}(\phi,v)$, and where

$$\Theta = \Theta(\phi) \equiv \beta + \rho\sigma_v i\phi, \tag{5.19}$$

and

$$\Lambda = \Lambda(\phi) \equiv i\phi - \phi^2. \tag{5.20}$$

The PDE (5.18) *is to be solved subject to the initial condition,*

$$\hat{H}(\phi,v,0) = e^{i\phi x_0}\delta(v-v_0). \tag{5.21}$$

Proof. By use of (5.17), and the assumptions that

$$\lim_{x\to\pm\infty} H(x,v,\tau) = \lim_{x\to\pm\infty} \frac{\partial H}{\partial x}(x,v,\tau) = 0.$$

We have

$$\mathcal{F}_x\left\{\frac{\partial H}{\partial x}\right\} = -i\phi\hat{H}, \quad \mathcal{F}_x\left\{\frac{\partial^2 H}{\partial x^2}\right\} = -\phi^2\hat{H},$$

$$\mathcal{F}_x\left\{\frac{\partial^2 H}{\partial x \partial v}\right\} = -i\phi\frac{\partial \hat{H}}{\partial v}, \quad \mathcal{F}_x\left\{\frac{\partial H}{\partial \tau}\right\} = \frac{\partial \hat{H}}{\partial \tau}.$$

Thus a direct application of the transform (5.17) to the PDE (5.15) for H yields the PDE (5.18) for $\hat{H}$. □

To reduce (5.18) to a first order PDE, we require the use of one more integral transform in the v dimension, namely the Laplace transform. The Laplace transform has been used in Feller (1951) to solve the Fokker–Planck equation associated with the diffusion process (5.2) for v. Here we use the same approach, the only difference being that we are now considering the joint transition density of the underlying asset driven by the mean-reverting process, v. By the nature of transition density functions, we make certain assumptions about the behavior of $\hat{H}(\phi,v,\tau)$. Specifically, we assume that $e^{-sv}\hat{H}(\phi,v,\tau)$ and $e^{-sv}\partial\hat{H}(\phi,v,\tau)/\partial v$ tend to zero as $v \to \infty$, where s

is the Laplace transform variable.[2] Proposition 5.3 provides the Laplace transform of (5.18) with respect to v.

Proposition 5.3. *Let $\mathcal{L}_v\{\hat{H}(\phi, v, \tau)\}$ be the Laplace transform of $\hat{H}(\phi, v, \tau)$ taken with respect to v, defined as*

$$\mathcal{L}_v\{\hat{H}(\phi, v, \tau)\} = \int_0^\infty e^{-sv}\hat{H}(\phi, v, \tau)dv \equiv \tilde{H}(\phi, s, \tau), \tag{5.22}$$

where the Laplace transform variable s could be complex whenever an improper integral exists. Applying the Laplace transform to (5.18) we find that $\tilde{H}$ satisfies the PDE,

$$\frac{\partial \tilde{H}}{\partial \tau} + \left(\frac{\sigma_v^2}{2}s^2 - \Theta s + \frac{\Lambda}{2}\right)\frac{\partial \tilde{H}}{\partial s} = \left[(\alpha - \sigma_v^2)s + \Theta - i\phi(r-q)\right]\tilde{H} + f_L(\tau), \tag{5.23}$$

with initial condition $\tilde{H}(\phi, s, 0) = e^{i\phi x_0 - sv_0}$, and where $f_L(\tau) \equiv (\sigma_v^2/2 - \alpha)\hat{H}(\phi, 0, \tau)$ is determined by the condition that

$$\lim_{s\to\infty} \tilde{H}(\phi, s, \tau) = 0. \tag{5.24}$$

Proof. Refer to Appendix 4.D in Chapter 4. The reader should keep in mind that in Chapter 4 we also have the jump term whereas, now we do not have the jump term and so the derivation is simpler. □

Equation (5.23) is now a first order PDE which can be solved using the method of characteristics. The unknown function $f_L(\tau)$ on the right hand side of (5.23) is then found by applying the condition (5.24). In this way we are able to solve (5.23) for $\tilde{H}(\phi, s, \tau)$, and the result is given in Proposition 5.4.

Proposition 5.4. *Using the method of characteristics, and subsequently applying condition (5.24) to determine $f_L(\tau)$, the solution to the first order*

[2]These seem quite reasonable assumptions to impose on a density function.

PDE (5.23) *is*

$$\tilde{H}(\phi, s, \tau) = \exp\left\{\left[\frac{(\alpha - \sigma_v^2)(\Theta - \Omega)}{\sigma_v^2} + \Theta - i\phi(r - q)\right]\tau\right\}$$
$$\times \left(\frac{2\Omega}{(\sigma_v^2 s - \Theta + \Omega)(e^{\Omega\tau} - 1) + 2\Omega}\right)^{2 - \frac{2\alpha}{\sigma_v^2}}$$
$$\times e^{i\phi x_0} e^{-\left(\frac{\Theta - \Omega}{\sigma_v^2}\right) v_0} \exp\left\{\frac{-2\Omega v_0 (\sigma_v^2 s - \Theta + \Omega) e^{\Omega\tau}}{\sigma_v^2[(\sigma_v^2 s - \Theta + \Omega)(e^{\Omega\tau} - 1) + 2\Omega]}\right\}$$
$$\times \Gamma\left(\frac{2\alpha}{\sigma_v^2} - 1; \frac{2\Omega v_0 e^{\Omega\tau}}{\sigma_v^2 (e^{\Omega\tau} - 1)} \times \frac{2\Omega}{(\sigma_v^2 s - \Theta + \Omega)(e^{\Omega\tau} - 1) + 2\Omega}\right), \tag{5.25}$$

where

$$\Omega = \Omega(\phi) \equiv \sqrt{\Theta^2(\phi) - \sigma_v^2 \Lambda(\phi)}, \tag{5.26}$$

$\Gamma(n; z)$ *is a (lower) incomplete gamma function, defined as*

$$\Gamma(n; z) = \frac{1}{\Gamma(n)} \int_0^z e^{-\xi} \xi^{n-1} d\xi, \tag{5.27}$$

and $\Gamma(n)$ *is the (complete) gamma function given by*

$$\Gamma(n) = \int_0^\infty e^{-\xi} \xi^{n-1} d\xi. \tag{5.28}$$

Proof. The reader should be advised that the proof is almost identical to that of Proposition 4.5. The only difference being that the jump term is not now included and leads to a slight re-definition of the term Λ. □

Having determined $\tilde{H}(\phi, s, \tau)$, we now seek to revert back to the original variables S and v, and thus obtain the density function $G(S, v, \tau)$. We begin this process by inverting the Laplace transform[3] in Proposition 5.5, again using the techniques of Feller (1951).

[3]The inverse Laplace transform is formally defined as

$$\mathcal{L}_v^{-1}\{\tilde{H}(\phi, s, \tau)\} = \hat{H}(\phi, v, \tau) = \frac{1}{2\pi i} \int_{c - i\infty}^{c + i\infty} e^{sv} \tilde{H}(\phi, s, \tau) ds,$$

where $c > 0$. For our purpose we do not need to evaluate the contour integral as the inverse Laplace transforms that we require are known and tabulated.

Proposition 5.5. *The inverse Laplace transform of $\tilde{H}(\phi, s, \tau)$ in* (5.25) *is*

$$\hat{H}(\phi, v, \tau) = \exp\left\{\frac{(\Theta - \Omega)}{\sigma_v^2}(v - v_0 + \alpha\tau)\right\} e^{i\phi x_0 - i\phi(r-q)\tau}$$
$$\times \frac{2\Omega e^{\Omega\tau}}{\sigma_v^2(e^{\Omega\tau} - 1)} \left(\frac{v_0 e^{\Omega\tau}}{v}\right)^{\frac{\alpha}{\sigma_v^2} - \frac{1}{2}} \exp\left\{\frac{-2\Omega}{\sigma_v^2(e^{\Omega\tau} - 1)}(v_0 e^{\Omega\tau} + v)\right\}$$
$$\times I_{\frac{2\alpha}{\sigma_v^2} - 1}\left(\frac{4\Omega}{\sigma_v^2(e^{\Omega\tau} - 1)}(v_0 v e^{\Omega\tau})^{\frac{1}{2}}\right), \tag{5.29}$$

where $I_k(z)$ is the modified Bessel function of the first kind, defined as

$$I_k(z) = \sum_{n=0}^{\infty} \frac{\left(\frac{z}{2}\right)^{2n+k}}{\Gamma(n + k + 1)n!}. \tag{5.30}$$

Proof. Refer to Appendix 4.F in Chapter 4. The reader should keep in mind that in Chapter 4 we also have the jump term whereas, now we do not have the jump term and so the derivation is simpler. □

All that remains now is to find the inverse Fourier transform of (5.29), and return to the original variable S and function $G(S, v, \tau; S_0, v_0)$. This last result is provided in Proposition 5.7.

Proposition 5.6. *Given the definition of the Fourier transform $\mathcal{F}_x$ in* (5.17), *the inverse Fourier transform of $\hat{H}(\phi, v, \tau)$ is*

$$\mathcal{F}_x^{-1}\{\hat{H}(\phi, v, \tau)\} = \frac{1}{2\pi}\int_{-\infty}^{\infty} e^{-i\phi x}\hat{H}(\phi, v, \tau)d\phi = H(x, v, \tau). \tag{5.31}$$

Using (5.31), *the inverse Fourier transform of* (5.29), *it can be shown that,*

$$H(x, v, \tau; x_0, v_0) = \frac{1}{2\pi}\int_{-\infty}^{\infty} e^{i\phi x_0} e^{\frac{(\Theta-\Omega)}{\sigma_v^2}(v - v_0 + \alpha\tau)} e^{-i\phi(r-q)\tau} e^{-i\phi x}$$
$$\times \frac{2\Omega e^{\Omega\tau}}{\sigma_v^2(e^{\Omega\tau} - 1)} \left(\frac{v_0 e^{\Omega\tau}}{v}\right)^{\frac{\alpha}{\sigma_v^2} - \frac{1}{2}}$$
$$\times \exp\left\{\frac{-2\Omega}{\sigma_v^2(e^{\Omega\tau} - 1)}(v_0 e^{\Omega\tau} + v)\right\}$$
$$\times I_{\frac{2\alpha}{\sigma_v^2} - 1}\left(\frac{4\Omega}{\sigma_v^2(e^{\Omega\tau} - 1)}(v_0 v e^{\Omega\tau})^{\frac{1}{2}}\right) d\phi. \tag{5.32}$$

Proof. The result follows by simply substituting (5.25) into (5.31) and rearranging. □

Now that we have managed to solve for $H(x, v, \tau)$, we now wish to revert back to the original density function and underlying asset variables.

Proposition 5.7. *The density function in terms of the original state variables can be represented as,*

$$
\begin{aligned}
G(S, v, \tau; S_0, v_0) = \frac{1}{2\pi} \int_{-\infty}^{\infty} & e^{i\phi \ln S_0} e^{\frac{(\Theta-\Omega)}{\sigma_v^2}(v - v_0 + \alpha\tau)} e^{-i\phi(r-q)\tau} e^{-i\phi \ln S} \\
& \times \frac{2\Omega e^{\Omega\tau}}{\sigma_v^2(e^{\Omega\tau} - 1)} \left(\frac{v_0 e^{\Omega\tau}}{v} \right)^{\frac{\alpha}{\sigma_v^2} - \frac{1}{2}} \\
& \times \exp\left\{ \frac{-2\Omega}{\sigma_v^2(e^{\Omega\tau} - 1)} (v_0 e^{\Omega\tau} + v) \right\} \\
& \times I_{\frac{2\alpha}{\sigma_v^2} - 1} \left(\frac{4\Omega}{\sigma_v^2(e^{\Omega\tau} - 1)} (v_0 v e^{\Omega\tau})^{\frac{1}{2}} \right) d\phi.
\end{aligned} \tag{5.33}
$$

Proof. Recall that $S = e^x$ and $H(x, v, \tau; S_0, v_0) = G(e^x, v, \tau; e^{x_0}, v_0)$. Substituting these into (5.32) we obtain the result in the above proposition. $\square$

With the knowledge of the density function and European option component, we can determine the early exercise premium by applying the Duhamel's principle.

5.4 Solution for the American Call Option

Having established the density function we can now carry out the calculations in the full representation of the American Call option value (5.13). The solution involves two linear components: the price of the corresponding European call option component on the underlying, plus the early exercise premium. We begin by finding the price of the European call option component.

The payoff for the European call is independent of v, this allows us to simplify equation (5.29) before taking the inverse Fourier transform to find the European call price, the result of which we present in Proposition 5.8.

Proposition 5.8. *Evaluating equation* (5.12) *when the payoff is independent of* v*, and applying the inverse Fourier transform* (5.31)*, the price of a European call option,* $C_E(S, v, \tau)$*, is given by*

$$
C_E(S, v, \tau) = S e^{-q\tau} P_1^H(S, v, \tau; K) - K e^{-r\tau} P_2^H(S, v, \tau; K), \tag{5.34}
$$

where

$$P_j^H(S, v, \tau; K) = \frac{1}{2} + \frac{1}{\pi} \int_0^\infty Re\left(\frac{f_j(S, v, \tau; \phi) e^{-i\phi \ln K}}{i\phi} \right) d\phi, \tag{5.35}$$

for $j = 1, 2$, *with*

$$\begin{aligned} f_j(S, v, \tau; \phi) &= \exp\{B_j(\phi, \tau) + D_j(\phi, \tau)v + i\phi \ln S\}, \\ B_j(\phi, \tau) &= i\phi(r - q)\tau + \frac{\alpha}{\sigma_v^2} \left\{ (\Theta_j + \Omega_j)\tau - 2 \ln \left(\frac{1 - Q_j e^{\Omega_j \tau}}{1 - Q_j} \right) \right\}, \\ D_j(\phi, \tau) &= \frac{(\Theta_j + \Omega_j)}{\sigma_v^2} \left(\frac{1 - e^{\Omega_j \tau}}{1 - Q_j e^{\Omega_j \tau}} \right), \end{aligned} \tag{5.36}$$

and $Q_j = (\Theta_j + \Omega_j)/(\Theta_j - \Omega_j)$, *where we define* $\Theta_1 = \Theta(i - \phi)$, $\Omega_1 = \Omega(i - \phi)$, *and* $\Theta_2 = \Theta(-\phi)$, $\Omega_2 = \Omega(-\phi)$.

Proof. Refer to Appendix 5.A. □

We note that (5.34) is the solution derived by Heston (1993) using characteristic functions. The details in Appendix 5.A demonstrate how to arrive at Heston's result by using the Fourier transform approach. In addition, we show that the general solution of Feller (1951) using Laplace transforms is consistent with Heston's solution. In particular, it is clear that one does not need to assume that the solution is of the form given by Heston (1993), but rather this can be derived directly using the Fourier and Laplace transform technique.

Proposition 5.9. *By using the Duhamel's principle, the early exercise premium for the American call,* $C_P(S, v, \tau)$, *is given by*

$$\begin{aligned} C_P(S, v, \tau) = \int_0^\tau \int_0^\infty & [qSe^{-q(\tau-\xi)} P_1^A(S, v, \tau - \xi; w, a(w, \xi)) \\ & - rKe^{-r(\tau-\xi)} P_2^A(S, v, \tau - \xi; w, a(w, \xi))] dw d\xi, \end{aligned} \tag{5.37}$$

where

$$\begin{aligned} & P_j^A(S, v, \tau - \xi; w, a(w, \xi)) \\ & \quad = \frac{1}{2} + \frac{1}{\pi} \int_0^\infty Re\left(\frac{g_j(S, v, \tau - \xi; \phi, w) e^{-i\phi \ln a(w, \xi)}}{i\phi} \right) d\phi, \end{aligned} \tag{5.38}$$

for $j = 1, 2$, *with*

$$g_j(S, v, \tau - \xi; \phi, w) = e^{\frac{(\Theta_j - \Omega_j)}{\sigma_v^2}(v - w + \alpha(\tau - \xi))} e^{i\phi(r-q)(\tau-\xi)} e^{i\phi \ln S}$$
$$\times \frac{2\Omega_j e^{\Omega_j(\tau-\xi)}}{\sigma_v^2(e^{\Omega_j(\tau-\xi)} - 1)} \left(\frac{w e^{\Omega_j(\tau-\xi)}}{v} \right)^{\frac{\alpha}{\sigma_v^2} - \frac{1}{2}}$$
$$\times \exp\left\{ \frac{-2\Omega_j}{\sigma_v^2(e^{\Omega_j(\tau-\xi)} - 1)} (w e^{\Omega_j(\tau-\xi)} + v) \right\}$$
$$\times I_{\frac{2\alpha}{\sigma_v^2} - 1} \left(\frac{4\Omega_j}{\sigma_v^2(e^{\Omega_j(\tau-\xi)} - 1)} (wve^{\Omega_j(\tau-\xi)})^{\frac{1}{2}} \right), \tag{5.39}$$

with $I_k(z)$ *given by* (5.30), *and* Θ_j *and* Ω_j *are given in Proposition* 5.8.

Proof. The reader should be advised that the proof is almost identical to that of Proposition 4.9. The only difference being that the jump term is included in Chapter 4 and leads to a slight re-definition of the term Λ. □

Having established the European call price and early exercise premium in Propositions 5.8 and 5.9, we can now obtain the integral expression for the American call price, along with the integral equation for the early exercise surface.

Proposition 5.10. *Substituting equations* (5.34) *and* (5.37) *into* (5.13), *the price of the American call option,* $C_A(S, v, \tau)$, *is given by the integral expression*

$$C_A(S, v, \tau) = Se^{-q\tau} P_1^H(S, v, \tau; K) - Ke^{-r\tau} P_2^H(S, v, \tau; K)$$
$$+ \int_0^\tau \int_0^\infty [qSe^{-q(\tau-\xi)} P_1^A(S, v, \tau - \xi; w, a(w, \xi))$$
$$- rKe^{-r(\tau-\xi)} P_2^A(S, v, \tau - \xi; w, a(w, \xi))] dw d\xi, \tag{5.40}$$

where P_j^H *and* P_j^A *are given by* (5.35) *and* (5.38) *respectively. Furthermore, evaluating* (5.40) *at* $S = a(v, \tau)$ *and using the boundary condition* (5.6), *we find that the integral equation satisfies,*

$$a(v, \tau) - K = a(v, \tau)e^{-q\tau} P_1^H(a(v, \tau), v, \tau; K) - Ke^{-r\tau} P_2^H(a(v, \tau), v, \tau; K)$$
$$+ \int_0^\tau \int_0^\infty [qa(v, \tau)e^{-q(\tau-\xi)} P_1^A(a(v, \tau), v, \tau - \xi; w, a(w, \xi))$$
$$- rKe^{-r(\tau-\xi)} P_2^A(a(v, \tau), v, \tau - \xi; w, a(w, \xi))] dw d\xi. \tag{5.41}$$

Proof. Direct substitution of (5.34) and (5.37) into (5.13) yields (5.40), and using (5.6) with direct substitution of $S = a(v, \tau)$ into (5.40) produces (5.41). □

Analytic representations for the sensitivities of C_A can be found by direct differentiation of (5.40). Here we provide the integral expression for the delta of $C_A(S, v, \tau)$.

Proposition 5.11. *Differentiating* (5.40) *with respect to* S*, the delta for the American call option, denoted by* $\Delta_A(S, v, \tau)$*, is given by*

$$
\begin{aligned}
&\Delta_A(S, v, \tau) \\
&\quad = e^{-q\tau} P_1^H(S, v, \tau; K) \\
&\qquad + \int_0^\tau \int_0^\infty qSe^{-q(\tau-\xi)} P_1^A(S, v, \tau-\xi; w, a(w,\xi)) dw d\xi \\
&\qquad + \int_0^\tau \int_0^\infty \Bigg\{ qe^{-q(\tau-\xi)} \left[\frac{1}{\pi} \int_0^\infty Re(e^{-i\phi \ln a(w,\xi)} g_1(S, v, \tau-\xi; \phi, w)) d\phi \right] \\
&\qquad - \frac{rK}{S} e^{-r(\tau-\xi)} \left[\frac{1}{\pi} \int_0^\infty Re(e^{-i\phi \ln a(w,\xi)} g_2(S, v, \tau-\xi; \phi, w)) d\phi \right] \Bigg\} dw d\xi,
\end{aligned}
\tag{5.42}
$$

where P_j^H *and* g_j *are given by (5.35) and (5.39) respectively.*

Proof. Refer to Appendix 5.C. □

In deriving (5.40) we have provided a generalisation of Heston's derivation for the European call price under the dynamics (2.10) and (2.11). Our approach focuses on application of the transforms directly to the PDE, and use of Duhamel's principle to obtain the early exercise premium. This generalises the solution of Heston to the American case, although without direct use of the characteristic function for the joint and marginal densities of S and v.

The solution (5.40)–(5.41) has also been derived by Tzavalis and Wang (2003) using the Snell envelope approach of Karatzas (1988). In their solution, Tzavalis and Wang use expectation operators to represent the early exercise premium, whereas here we provide explicit expressions for these expectations over the joint density of S and v. We also derive the joint density and rewrite these expectations in a form that is analogous to the Heston (1993) solution for the European call.

Equations (5.40) and (5.41) are presented in a form that matches the American call price presented by Kim (1990) and Jamshidian (1992) in the constant volatility case. Thus (5.40) is a generalisation of the early exercise premium decomposition of an American call option to incorporate Heston's stochastic volatility dynamics. Furthermore, we see that (5.40) is an integral expression for the American call price that depends upon the unknown early exercise surface $a(v, \tau)$. As in the constant volatility case, we can readily price the American call using (5.40) with numerical integration once $a(v, \tau)$ has been determined.

The free surface satisfies the two-dimensional Volterra integral equation (5.41). Since $0 \leq v < \infty$, one must find a way to efficiently estimate the boundary for all values of v. Once this has been established, standard techniques can be used to solve (5.41) numerically. Hence the two-pass process observed in the constant volatility case, where one first finds $a(v, \tau)$ and then $C_A(S, v, \tau)$, is still valid for the stochastic volatility case. The remainder of this chapter will be concerned with finding an efficient scheme to solve (5.41) for the free boundary.

5.5 Numerical Scheme for the Free Surface

The integral equation (5.41) for $a(v, \tau)$ must be solved numerically. One approach is to discretise the integrals with respect to the running time-to-maturity variable ξ and the running volatility variable w using suitable quadrature schemes. If we use an m-point scheme for the integral with respect to w, we will have to solve a system of m equations for $a(v, \tau)$ at each increasing value of τ. In practice this is a very cumbersome approach for several reasons. The terms P^H and, in particular, P^A must be evaluated numerically as well. This means that the early exercise premium involves a three-dimensional integration. Furthermore, as the number of time steps is increased, the computational burden involved in solving (5.41) increases at a "faster than linear" rate. In light of these complexities, it is highly desirable to find ways to reduce the amount of numerical integration required.

A useful approach to reducing the computational burden in solving the integral equation is provided by Tzavalis and Wang (2003). According to the empirical findings of Broadie *et al.* (2000), there is evidence that $\ln a(v, \tau)$ is well approximated by a function that is linear in v. On this basis Tzavalis and Wang (2003) expand $\ln a(v, \tau)$ in a Taylor series about θ (the long-run volatility) so that $a(v, \tau)$ can be approximated according to

$$\ln a(v, \tau) \approx a_0(\tau) + a_1(\tau)v, \tag{5.43}$$

where a_0 and a_1 are functions of τ that remain to be determined. Substituting (5.43) into (5.40), we derive a simplified system of integral equations for $C_A(S, v, \tau)$ and $a(v, \tau)$.

Proposition 5.12. *Under the linear approximation for* $\ln a(v, \tau)$ *given by* (5.43), *the price of an American call option written on* S, *may be expressed as*

$$C_A(S,v,\tau) = Se^{-q\tau}P_1^H(S,v,\tau;K) - Ke^{-r\tau}P_2^H(S,v,\tau;K) + \int_0^\tau qSe^{-q(\tau-\xi)}\bar{P}_1^A(S,v,\tau-\xi;e^{a_0(\xi)},e^{a_1(\xi)})d\xi - \int_0^\tau rKe^{-r(\tau-\xi)}\bar{P}_2^A(S,v,\tau-\xi;e^{a_0(\xi)},e^{a_1(\xi)})d\xi, \quad (5.44)$$

where

$$\bar{P}_j^A(S,v,\tau;e^{a_0(\xi)},a_1(\xi)) \equiv \frac{1}{2} + \frac{1}{\pi}\int_0^\infty Re\left(\frac{e^{-i\phi a_0(\xi)}}{i\phi}\bar{f}_j(S,v,\tau;\phi,-a_1(\xi)\phi)\right)d\phi, \quad (5.45)$$

for $j = 1, 2$. *The function* $\bar{f}_j(S, v, \tau; \phi, \psi)$ *is given by*

$$\bar{f}_j(S,v,\tau;\phi,\psi) = \exp\{\bar{B}_j(\phi,\psi,\tau) + \bar{D}_j(\phi,\psi,\tau)v + i\phi\ln S\},$$

$$\bar{B}_j(\phi,\psi,\tau) = i\phi(r-q)\tau + \frac{\alpha}{\sigma_v^2}\left\{(\Theta_j+\Omega_j)\tau - 2\ln\left(\frac{1-\bar{Q}_je^{\Omega_j\tau}}{1-\bar{Q}_j}\right)\right\},$$

$$\bar{D}_j(\phi,\psi,\tau) = i\psi + \frac{(\Theta_j - \sigma_v^2 i\psi + \Omega_j)}{\sigma_v^2}\left(\frac{1-e^{\Omega_j\tau}}{1-\bar{Q}_je^{\Omega_j\tau}}\right), \quad (5.46)$$

with $\bar{Q}_j = (\Theta_j - \sigma_v^2 i\psi + \Omega_j)/(\Theta_j - \sigma_v^2 i\psi - \Omega_j)$. *The terms* Θ_j *and* Ω_j *are as defined in Proposition* 5.8.

Furthermore, by use of the boundary condition (5.6), the free surface satisfies the integral equation

$$e^{a_0(\tau)+a_1(\tau)v} - K = e^{a_0(\tau)+a_1(\tau)v}e^{-q\tau}P_1^H(e^{a_0(\tau)+a_1(\tau)v},v,\tau;K) - Ke^{-r\tau}P_2^H(e^{a_0(\tau)+a_1(\tau)v},v,\tau;K) + \int_0^\tau qe^{a_0(\tau)+a_1(\tau)v}e^{-q(\tau-\xi)}\bar{P}_1^A(e^{a_0(\tau)+a_1(\tau)v},v,\tau-\xi;e^{a_0(\xi)},e^{a_1(\xi)})d\xi - \int_0^\tau rKe^{-r(\tau-\xi)}\bar{P}_2^A(e^{a_0(\tau)+a_1(\tau)v},v,\tau-\xi;e^{a_0(\xi)},e^{a_1(\xi)})d\xi. \quad (5.47)$$

Proof. Refer to Appendix 5.D. □

Comparing equation (5.44) with Kim's 1990 solution, we can readily draw parallels between the cumulative normal density functions appearing in the constant volatility case, and the $\bar{P}_j^A$ terms occurring in the stochastic volatility case, here expressed as inverse Fourier transforms. It is also interesting to note that the $\bar{P}_j^A$ terms are related to the P_j^H of the Heston solution via $\bar{P}_j^A(S, v, \tau; K, 0) = P_j^H(S, v, \tau; K)$.

The key advantage that arises from the use of the assumption (5.43) is that (5.44) and (5.47) involve double integral terms, rather than the three dimensional integrals arising in (5.40) and (5.41). This makes it far more appealing to solve (5.47) for the functions $a_0(\tau)$ and $a_1(\tau)$ rather than solving (5.41) for the more general function $a(v, \tau)$.

We require a system of two equations to determine a_0 and a_1. These are established by evaluating (5.47) at two distinct values of v, namely $v_0(\tau)$ and $v_1(\tau)$ the choice of which will be discussed below, producing the system

$$C_A(a(v_0(\tau), \tau), v, \tau) = e^{a_0(\tau)+v_0(\tau)a_1(\tau)} - K, \tag{5.48}$$

$$C_A(a(v_1(\tau), \tau), v, \tau) = e^{a_0(\tau)+v_1(\tau)a_1(\tau)} - K, \tag{5.49}$$

which may also be written as

$$a_1(\tau) = \frac{1}{v_0(\tau)} \left(\ln \left[C_A(\exp(a_0(\tau) + v_0(T)a_1(\tau), v_0(\tau), \tau)) + K\right] - a_0(\tau)\right), \tag{5.50}$$

$$a_0(\tau) = \ln \left[C_A(\exp(a_0(\tau) + v_1(\tau)a_1(\tau), v_1(\tau), \tau)) + K\right] - v_1(\tau)a_1(\tau). \tag{5.51}$$

Tzavalis and Wang (2003) solve (5.48), (5.49) separately using Chebyshev polynomials. They then calculate $a_0(\tau)$ and $a_1(\tau)$ using the approximating boundary function (5.43).

Here we propose to solve (5.50), (5.51) using an iterative method based on numerical techniques frequently applied to Volterra integral equations. We discretise the time domain into N subintervals of length $\Delta\tau$, such that $T = N\Delta\tau$. At step $n = 0$ we know from Kim (1990) that the early exercise boundary is

$$a(v, 0) = \max\left(\frac{r}{q}K,\ K\right), \tag{5.52}$$

a result which also holds true under stochastic volatility, since the payoff for the American call does not explicitly depend upon v. It follows that for the linear approximation of the free boundary

$$a_0(0) = \max\left(\ln\left[\frac{r}{q}K\right],\ \ln K\right),\quad a_1(0) = 0. \tag{5.53}$$

This provides us with the initial values of a_0 and a_1 for the time-stepping procedure.

At each subsequent time step, $n = 1, 2, \ldots, N$ we must determine two unknowns, namely $a_0^n = a_0(n\Delta\tau)$ and $a_1^n = a_1(n\Delta\tau)$, whose values depend upon each other, as well as on previous values of $a_0(n\Delta\tau)$ and $a_1(n\Delta\tau)$ (that is, $n = 0, 1, 2, \ldots, N-1$). When iterating for a_0^n and a_1^n we take as our initial approximations $a_{0,0}^n = a_0^{n-1}$, and $a_{1,0}^n = a_1^{n-1}$. We then solve the linked system of integral equations (5.48), (5.49) for the successive iterates $(a_{0,k}^n, a_{1,k}^n)$ according to

$$a_{1,k}^n = \frac{1}{v_0^n}\left(\ln\left[C_A(\exp(a_{0,k-1}^n + v_0^n a_{1,k}^n), v_0^n, \tau) + K\right] - a_{0,k-1}^n\right) \tag{5.54}$$

$$a_{0,k}^n = \ln\left[C_A(\exp(a_{0,k}^n + v_1^n a_{1,k}^n), v_1^n, \tau) + K\right] - v_1^n a_{1,k}^n, \tag{5.55}$$

where $v_0^n = v_0(n\Delta\tau)$ and $v_1^n = v_1(n\Delta\tau)$. We continue sequentially, solving (5.54)–(5.55) for $k = 1, 2, \ldots$ until both $|a_{0,k}^n - a_{0,k-1}^n| < \epsilon_0$ and $|a_{1,k}^n - a_{1,k-1}^n| < \epsilon_1$, for specified values of ϵ_0 and ϵ_1. Once the solutions have converged to the desired level of tolerance, the method advances to the next time step.

Since we are approximating the true early exercise boundary with a linear function of v, we seek values of v_0 and v_1 that will maximise the accuracy of this estimate. Tzavalis and Wang (2003) suggest using $\mathbb{E}_T[v(0)]$ and $\mathbb{E}_T[v(\tau)]$, since these values represent expected values of v over the time intervals applicable to the early exercise premium.[4] Dufresne (2001) shows that this expectation evaluates to

$$\mathbb{E}_T[v(\tau)] = \theta + (v(T) - \theta)e^{-\kappa_v(T-\tau)}. \tag{5.56}$$

The main disadvantage of this specification for v_0 and v_1 is that when $v(T)$ is close to θ, or τ is close to T, the values of v_0 and v_1 will be nearly identical. This creates difficulties when trying to solve the system (5.54)–(5.55) since it tends towards being ill-conditioned. Here we consider an alternative specification that avoids this problem.

[4]Recall that τ is time remaining until maturity. Hence $\mathbb{E}_T[v(0)]$ represents the expected value of the volatility at the expiry date, taken at time T until maturity.

Using the results from Dufresne (2001), we can determine the long-run standard deviation for v. We then define v_0 and v_1 as[5]

$$v_0(\tau) = \mathbb{E}_T[v(\tau)] + \frac{\sigma_v}{|\kappa_v|}\sqrt{\frac{\kappa_v\theta}{2}}, \tag{5.57}$$

$$v_1(\tau) = \mathbb{E}_T[v(\tau)] - \frac{\sigma_v}{|\kappa_v|}\sqrt{\frac{\kappa_v\theta}{2}}. \tag{5.58}$$

This specification allows the a_1 term to capture the majority of the volatility dependence in the free boundary before estimating a_0, and ensures that $v_0(\tau)$ and $v_1(\tau)$ will always be distinct, forming a bounded interval around the expected value of the volatility at time to maturity τ, given the observed spot volatility $v(T)$. Figure 5.1 demonstrates this bound in the case where $v(T) > \theta$.

It is important to note that under this method a different $a_0(\tau)$ and $a_1(\tau)$ will be found for each value of $v(T)$. We emphasize that under the current time-to-maturity notation this would be the current value of v, which would be chosen from market considerations.[6] As such the approximation of Tzavalis and Wang (2003), and variations such as the one presented here, are designed to be applied for a single spot value of the volatility v.

Since the equations (5.54) and (5.55) for a_1 and a_0 are highly nonlinear, they must be solved using numerical techniques. We use the Newton's method for root finding,[7] and the time integrals in (5.44) are approximated using the compound trapezoidal rule that was found to be effective in solving the corresponding integral equation in the stochastic volatility case by Kallast and Kivinukk (2003). While the order of accuracy for the trapezoidal rule is not very large, the advantage of this quadrature scheme is that the weights for the method do not vary for even and odd numbers of integration points. Since we add one additional point to the integration scheme at each increasing time step, the compound trapezoidal rule provides a free boundary estimate that increases monotonically as time to

[5]Here we have used the result from Dufresne (2001) that

$$\mathrm{var}_T[v(\tau)] = \frac{\Theta\sigma_v^2}{2\kappa_v} - \left(\frac{\theta\sigma_v^2}{\kappa_v} - \frac{\sigma_v^2}{\kappa_v}v_0\right)e^{-\kappa_v\tau} + \left(\frac{\theta\sigma_v^2}{2\kappa_v} - \frac{\sigma_v^2}{\kappa_v}v_0\right)e^{-2\kappa_v\tau}$$

and in (5.57), (5.58) we take the maximum standard deviation band obtained by letting $(T-\tau) \to \infty$. See appendix 5.E.

[6]As a practical matter the value of $v(T)$ one would choose might best be the B-S implied volatility.

[7]Appendix 5.C gives the details of the implementation of Newton's method.

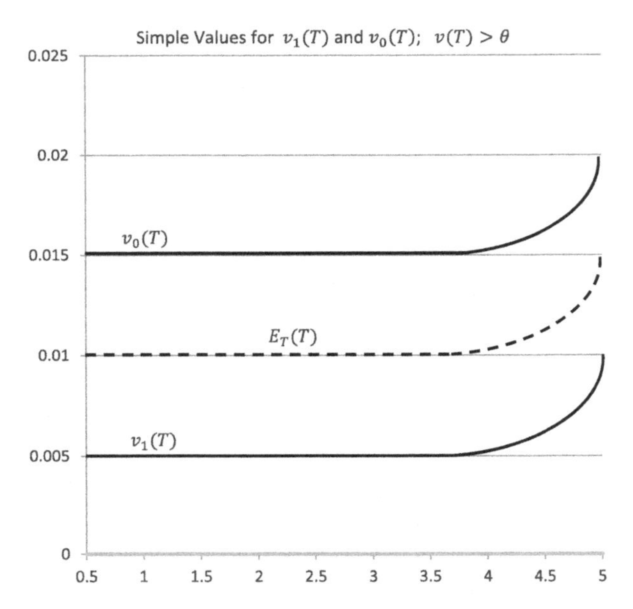

Fig. 5.1 Example of the values of v_0 and v_1 used to solve for $a_0(\tau)$ and $a_1(\tau)$, in the case where $v(T) > \theta$. Parameter values used are $T = 5$, $\theta = 0.01$, $\kappa_v = 2$, $\sigma_v = 0.10$ and $v(T) = 0.015$.

maturity increases. The terms P^H and $\bar{P}^A$ in equation (5.44), which involve complex integration, are evaluated using the methods of Kahl and Jäckel (2006).

We test the accuracy of the approximation (5.43) by benchmarking the method against two alternative numerical solutions applied directly to the PDE (5.3). The first is a typical projected successive overrelaxation (PSOR) method used in conjunction with a Crank-Nicolson finite difference scheme. The other benchmark is the method of lines approach, as detailed by Meyer and van der Hoek (1997).[8]

When exploring the convergence properties we compute the option prices and option deltas for four different cases that cover the cases $r > q$ and $r < q$ as well as positive and negative correlation;

- $r = 0.05$, $q = 0.03$ and $\rho = 0.5$,
- $r = 0.05$, $q = 0.03$ and $\rho = -0.5$,
- $r = 0.03$, $q = 0.05$ and $\rho = 0.5$,
- $r = 0.03$, $q = 0.05$ and $\rho = -0.5$.

[8]See appendix 5.F for details on the implementation of method of lines to the PDE (5.3).

Table 5.1 Runtime and Average RMSE for Numerical integration.

Time steps	Average Run time	Average Price RMSE	Average Delta RMSE	Average RRMSE
1	0.33	0.03958	0.00762	0.06379
10	6.55	0.00201	0.00104	0.00409
20	19.73	0.00089	0.00057	0.00178
30	39.75	0.00058	0.00040	0.00112
40	62.47	0.00045	0.00031	0.00082
50	93.59	0.00038	0.00026	0.00065
60	127.06	0.00034	0.00023	0.00055
70	165.41	0.00031	0.00020	0.00047
80	211.85	0.00030	0.00018	0.00042
90	257.35	0.00029	0.00017	0.00038
100	311.15	0.00028	0.00015	0.00035

Table 5.2 Runtimes and Average RMSE for PSOR.

Share Steps	Var Steps	Time Steps	Average Runtime	Average Price RMSE	Average Delta RMSE	Average RRMSE
80	80	320	2.38	0.06026	0.00240	0.11236
120	120	480	8.69	0.02695	0.00117	0.04979
160	160	640	19.61	0.01527	0.00074	0.02799
200	200	800	35.37	0.00931	0.00054	0.01785
240	240	960	58.42	0.00589	0.00039	0.01250
280	280	1120	86.51	0.00383	0.00030	0.00963
320	320	1280	124.97	0.00260	0.00024	0.00785
360	360	1440	175.72	0.00183	0.00019	0.00664
400	400	1600	238.76	0.00131	0.00015	0.00579

All other parameters are listed in Table 5.4. For each case we compute the RMSE and the Relative RMSE (RRMSE) for the set of option prices and option deltas over a mesh at the variances 0.03, 0.04 and 0.05 and at the share prices 80, 90, 100, 100 and 120. From this we calculate the Average RMSE and the average RRMSE by combining the four cases. The results from the three methods are summarized in Tables 5.1, 5.2 and 5.3 and displayed graphically in Figures 5.2 and 5.3 using efficiency plots. The reference solution was computed using PSOR to a high order of discretisation with stepsizes $\Delta S = 0.1$, $\Delta v = 0.0005$ and $\Delta t = 0.000025$.

The efficiency plots in Figure 5.2 compare the average RMSE versus average runtime to calculate the prices and deltas for the set of options

Table 5.3 Runtimes and Average RMSE for MOL.

Vol Steps	Time Steps	Share Steps	Average Runtime	Average Price RMSE	Average Delta RMSE	Average RRMSE
50	100	500	2.50	0.01299	0.00215	0.08269
75	150	750	10.00	0.00859	0.00136	0.05904
100	200	1000	21.63	0.00648	0.00107	0.04614
125	250	1250	48.00	0.00527	0.00089	0.03847
150	300	1500	70.76	0.00432	0.00071	0.03165
175	350	1750	111.38	0.00371	0.00061	0.02738
200	400	2000	159.30	0.00326	0.00055	0.02408
225	450	2250	229.21	0.00291	0.00049	0.02146
250	500	2500	313.44	0.00263	0.00043	0.01933

Table 5.4 Default parameter values used when pricing the American call under Heston's stochastic volatility model.

Parameter	Value	Parameter	Value
T	0.25	$v(T)$	0.04
r	0.05	θ	0.09
q	0.03	κ	4.00
K	100	σ	0.10
ρ	0	λ	0

using the three methods. In Figure 5.3 we compare the average runtime to calculate the RRMSE of the option prices plus that of the deltas to some given accuracy. This would more likely be the calculation of interest to practitioners who would require not only the option price but also the hedge quantity to the same relative accuracy. The points on the efficiency plots correspond to the different discretisations given in Tables 5.1, 5.2 and 5.3. Looking first at Figure 5.2 we see that as far as the calculation of the price is concerned the integration method is the most efficient (in terms of speed) at any given level of accuracy, the method of lines is initially more efficient than PSOR but eventually PSOR becomes more efficient. Perhaps this is not surprising since PSOR at a high discretisation provides the benchmark so that as the discretisation becomes higher it must tend to dominate the other methods. As far as the calculation of delta is concerned there is not much difference in the efficiency of the three methods though the MOL is slightly less efficient.

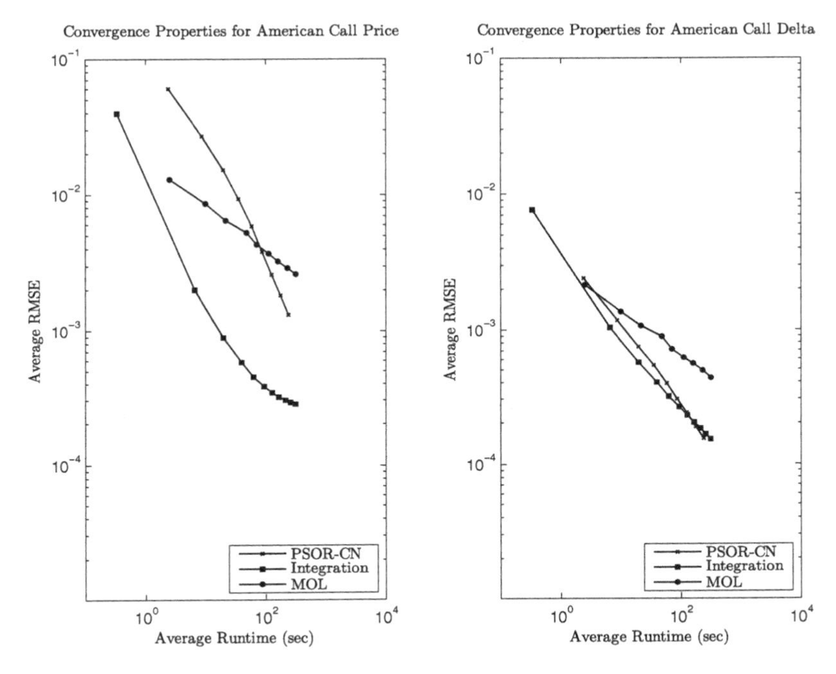

Fig. 5.2 Runtime efficiency for the American call price (left) and American call delta (right) for numerical integration, method of lines and PSOR.

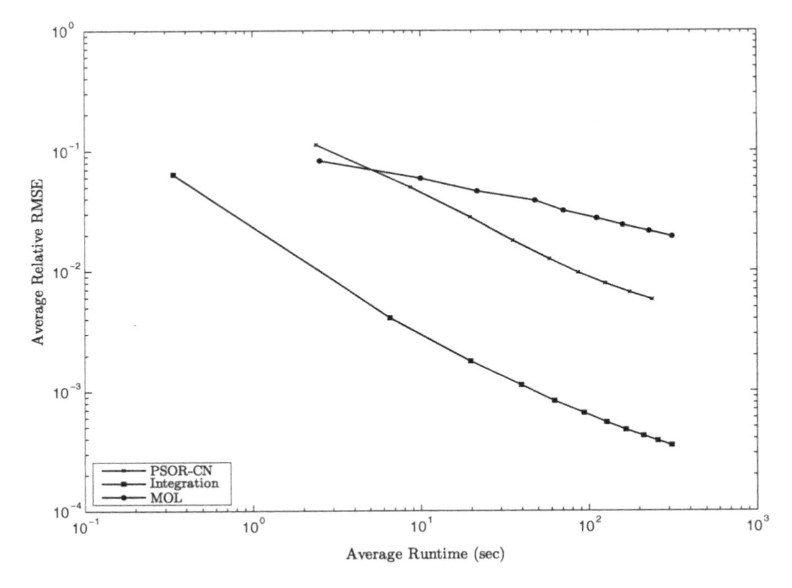

Fig. 5.3 Relative Runtime efficiency for the American call price plus the American call delta for numerical integration, method of lines and PSOR.

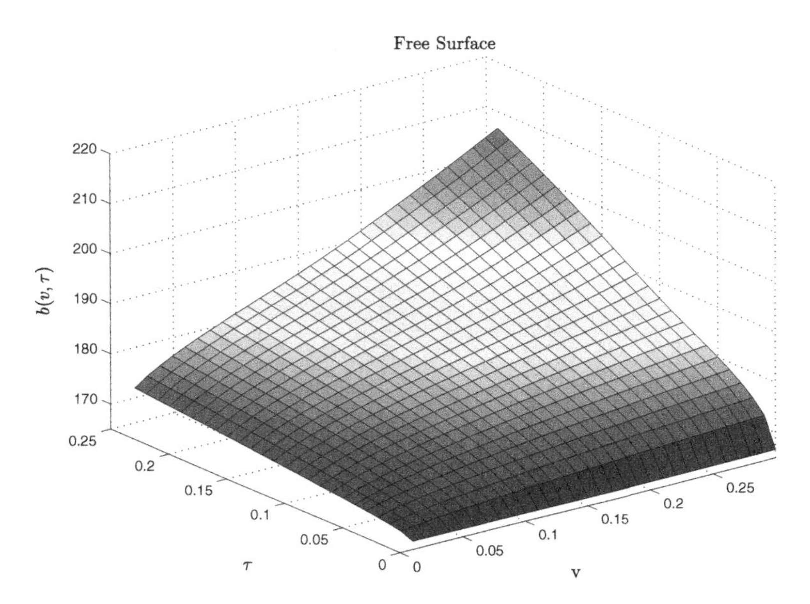

Fig. 5.4 The early exercise surface for an American call option when $r = 0.05$, $q = 0.03$ and $\rho = -0.5$; all other parameters are listed in Table 5.4.

When we come to compare the efficiency plots in terms of the RRMSE for the price plus the delta (so we calculate the sum of the difference of the true price plus the true delta to a given accuracy) in Figure 5.3 the integration method becomes clearly superior, this is due to the fact that in calculating the price we also calculate the delta (in order to apply Newton's method to solve (5.54) and (5.55) — also the calculation of the expressions in the delta (see Appendix 5.C) does not require a great deal of additional calculation since they mostly involve terms already calculated in generating the price.

Figure 5.4 displays a free surface calculated using the integration method. Figure 5.5 displays the relative differences in the free boundary values between PSOR and numerical integration across a (τ, v) grid. Since all the calculations require specification of an initial value of v (here denoted $v(T)$ since the argument of v is time-to-maturity) we have used two such values in Figure 5.5 ($v(T) = 0.04$ corresponding to 20% volatility and $v(T) = 0.20$ corresponding to 44.72% volatility) in order to gauge the difference on the free surface. The relative differences across the grid are not great indicating that the approximation (5.43) to the free boundary is quite reasonable. It could of course be improved by going to the quadratic term,

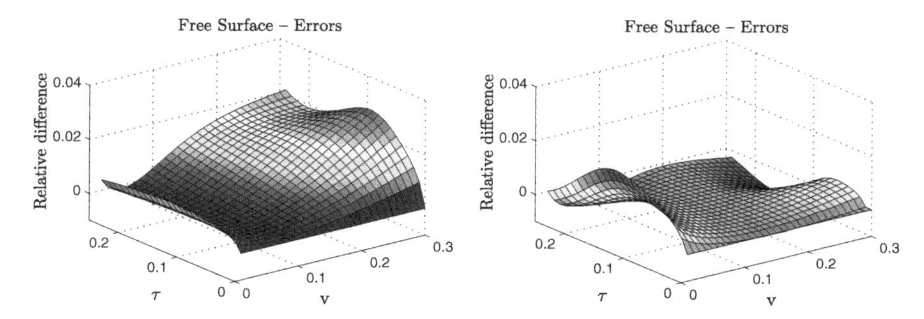

Fig. 5.5 Relative differences in the free boundary values between PSOR and numerical integration when $v(T) = 0.04$(left) and $v(T) = 0.20$(right). The maximum relative difference is 0.0279 and the minimum relative difference 0.0000539. (Right) The maximum relative difference is 0.0211 and the minimum relative difference −0.00824.

but this hardly seems necessary since the integration method is already so efficient in calculating the price and delta. It is already fairly well known in the case of American option valuation under constant volatility that it is not necessary to obtain a highly accurate approximation to the free boundary to obtain a good approximation to the price, we seem to be observing the some phenomenon in the stochastic volatility case.

5.6 Conclusion

This chapter has presented an analysis of the American call option pricing problem under stochastic volatility. Taking Heston's 1993 square root model for the random evolution of the asset volatility, Lewis (2000) and Detemple and Tian (2002) show that the option price is the solution to a two-dimensional free boundary value problem. Following the techniques of Jamshidian (1992), we show that the homogeneous partial differential equation (PDE) for the option price that needs to be solved in a restricted domain is replaced by an equivalent inhomogeneous PDE which has to be solved in an unrestricted domain. We then present the general solution of the resulting PDE by using Duhamel's principle. The general solution involves the transition density function, which rather than the transition density function satisfies the Kolmogorov PDE associated with the driving stochastic processes. We then present a natural, systematic approach to solving this PDE. Once solved, the full integral representation of the American option price is presented.

In general it is difficult to solve the integral equation for the free surface, as is involves three-dimensional integrals, and discretisation produces a system of integral equations corresponding to each discrete value of the volatility. To overcome this problem, we approximate the free surface by a form that is log-linear in the stochastic variance as suggested by Tzavalis and Wang (2003). This leads to a system of two integral equations for the free surface, which involve a two-dimensional integral. We develop a numerical algorithm for solving this integral equation system that basically uses standard ideas for the solution of Volterra integral equations. We benchmark the accuracy of the algorithm against finite differences, PSOR and the method of lines solutions. The accuracy of the three methods are compared using efficiency plots which compare the computational time versus accuracy trade-off for a set of option prices across a range of underlying stock prices. The algorithm for the numerical solution of the integral equations performs quite well if one wishes to calculate just the price and is reasonable if one wants to calculate the American option delta. The method comes into its own if one is interested in calculating both the price and the delta at the same time, which would probably be the situation most relevant to practitioners.

The general approach adopted here of transforming the American option pricing homogeneous free boundary value problem to an inhomogeneous boundary problem on an unrestricted region and then solving this by use of Duhamel's principle and transform methods can be extended to a number of interesting American option pricing problem. The technique could be extended to certain types of American exotic options under stochastic volatility such as basket options and compound options.

Appendix

5.A Proof of Proposition 5.8 — The European Option Price

We note that the payoff the European call option, $(e^x - K)^+$, is independent of v. This allows us to write the European option value as

$$C_E(S, v, \tau) = \frac{e^{-r\tau}}{2\pi} \int_{-\infty}^{\infty} \left(\int_{\ln K}^{\infty} (e^y - K) e^{i\phi y} dy \right) \times \left[\int_0^{\infty} U(x, v, \tau; -\phi, y, w) dw \right] d\phi dy. \tag{5.59}$$

We denote the integral with respect to w as $g_2(x, v, \tau)$ and set $f_2(S, v, t) = g_2(\ln S, v, t)$. Applying the transformations given by (4.77) and (4.80) in

Chapter 4 we can write

$$f_2(S,v,\tau) = e^{\ln S + i\phi(r-q)\tau} \exp\left\{\left[\frac{\alpha(\Theta-\Omega)}{\sigma^2}\right]\tau\right\}$$
$$\times \int_0^\infty e^{-A-y} \exp\left\{-\left(\frac{\Theta-\Omega}{\sigma^2}\right)\frac{\sigma^2(1-e^{-\Omega\tau})}{2\Omega}A\right\}$$
$$\times \left(\frac{A}{y}\right)^{\frac{\alpha}{\sigma^2}-\frac{1}{2}} \exp\left\{\frac{(\Theta-\Omega)(e^{\Omega\tau}-1)}{2\Omega}y\right\} I_{\frac{2\alpha}{\sigma^2}-1}(2\sqrt{Ay})dA. \tag{5.60}$$

To undertake the integration with respect to A, we write (5.60) as,

$$f_2(S,v,\tau) = e^{\ln S + i\phi(r-q)\tau} \exp\left\{\left[\frac{(\alpha-\sigma^2)(\Theta-\Omega)}{\sigma^2} + \Theta - \Omega\right]\tau\right\}$$
$$\times e^{-y} \exp\left\{\frac{(\Theta-\Omega)(e^{\Omega\tau}-1)y}{2\Omega}\right\} J(y),$$

where, by use of (5.30),

$$J(y) = \sum_{n=0}^{\infty} y^n \int_0^\infty \exp\left\{-\left[\left(\frac{\Theta-\Omega}{2\Omega}\right)(1-e^{-\Omega\tau})+1\right]A\right\}$$
$$\times \frac{A^{n+\frac{2\alpha}{\sigma^2}-1}}{n!\Gamma\left(\frac{2\alpha}{\sigma^2}+n\right)}dA.$$

Making the change of variable $\xi = [(\Theta-\Omega)(1-e^{-\Omega\tau})/2\Omega+1]A$, and using the definition of the gamma function from (5.28), we have

$$J(y) = \sum_{n=0}^{\infty} \frac{y^n}{n!\Gamma\left(\frac{2\alpha}{\sigma^2}+n\right)} \frac{2\Omega}{(\Theta-\Omega)(1-e^{-\Omega\tau})+2\Omega}$$
$$\times \int_0^\infty e^{-\xi}\left(\frac{2\Omega}{(\Theta-\Omega)(1-e^{-\Omega\tau})+2\Omega}\right)^{n+\frac{2\alpha}{\sigma^2}-1} \xi^{n+\frac{2\alpha}{\sigma^2}-1}d\xi$$
$$= \sum_{n=0}^{\infty} \frac{y^n}{n!}\left(\frac{2\Omega}{(\Theta-\Omega)(1-e^{-\Omega\tau})+2\Omega}\right)^{n+\frac{2\alpha}{\sigma^2}} \frac{\Gamma\left(\frac{2\alpha}{\sigma^2}+n\right)}{\Gamma\left(\frac{2\alpha}{\sigma^2}+n\right)}.$$

Recalling the Taylor expansion for e^x, we can express $J(y)$ as

$$J(y) = \left(\frac{2\Omega}{(\Theta-\Omega)(1-e^{-\Omega\tau})+2\Omega}\right)^{\frac{2\alpha}{\sigma^2}} \exp\left\{\frac{2\Omega y}{(\Theta-\Omega)(1-e^{-\Omega\tau})+2\Omega}\right\},$$

and thus $g(x, v, \tau)$ becomes

$$f_2(S, v, \tau) = e^{\ln S + i\phi(r-q)\tau} e^{\frac{(\Theta-\Omega)}{\sigma^2}(\alpha\tau+v)} \left(\frac{2\Omega}{(\Theta-\Omega)(1-e^{-\Omega\tau})+2\Omega} \right)^{\frac{2\alpha}{\sigma^2}}$$
$$\times \exp\left\{ \frac{-2\Omega v}{\sigma^2(e^{\Omega\tau}-1)} \right\}$$
$$\times \exp\left\{ \frac{2\Omega}{(\Theta-\Omega)(1-e^{-\Omega\tau})+2\Omega} \times \frac{2\Omega v}{\sigma^2(e^{\Omega\tau}-1)} \right\}. \quad (5.61)$$

We now take the inverse Fourier transform of (5.61), and return to the original variables. Before we apply the inverse transform, we will rewrite (5.61) in the form presented by [Heston (1993)] for ease of comparison.

We begin by writing $f_2(x, v, \tau)$ in the form

$$f_2(S, v, \tau; -\phi) = \exp\{-i\phi \ln S + B_2(-\phi, \tau) + D_2(-\phi, \tau)v\}, \quad (5.62)$$

where $B(\phi, \tau)$ and $D(\phi, \tau)$ are determined by equating (5.61) with (5.62). For $B_2(\phi, \tau)$ we have

$$B_2(\phi, \tau) = i\phi(r-q)\tau$$
$$+ \frac{\alpha}{\sigma^2} \left\{ (\Theta_2 - \Omega_2)\tau - 2\ln\left(\frac{(\Theta_2-\Omega_2)(1-e^{-\Omega_2\tau})+2\Omega_2}{2\Omega_2} \right) \right\},$$

where $\Theta_2 = \Theta_2(\phi) \equiv \Theta(-\phi)$ and $\Omega_2 = \Omega_2(\phi) \equiv \Omega(-\phi)$. Defining $Q_2 = Q_2(\phi) \equiv (\Theta_2 + \Omega_2)/(\Theta_2 - \Omega_2)$, the expression for $B_2(\phi, \tau)$ thus becomes

$$B_2(\phi, \tau) = i\phi(r-q)\tau + \frac{\alpha}{\sigma^2} \left\{ (\Theta_2 + \Omega_2)\tau - 2\ln\left(\frac{1 - Q_2 e^{\Omega_2\tau}}{1 - Q_2} \right) \right\}. \quad (5.63)$$

For $D_2(\phi, \tau)$ we find that

$$D_2(\phi, \tau) = \frac{(\Theta_2 - \Omega_2)}{\sigma^2} - \frac{2\Omega_2}{\sigma^2(e^{\Omega_2\tau}-1)}$$
$$+ \frac{2\Omega_2}{(\Theta_2-\Omega_2)(1-e^{-\Omega_2\tau})+2\Omega_2} \times \frac{2\Omega_2}{\sigma^2(e^{\Omega_2\tau}-1)},$$

which simplifies to

$$D_2(\phi, \tau) = \frac{(\Theta_2 + \Omega_2)}{\sigma^2} \left[\frac{1 - e^{\Omega_2\tau}}{1 - Q_2 e^{\Omega_2\tau}} \right]. \quad (5.64)$$

Taking the inverse Fourier transform of (5.62), and returning to the variable S, we find that

$$C_E(S,v,\tau) = \frac{e^{-r\tau}}{2\pi}\int_{-\infty}^{\infty}\left(\int_{\ln K}^{\infty}(e^y - K)e^{i\phi y}dy\right)\exp\{B_2(-\phi,\tau)$$

$$+ D_2(-\phi,\tau)v\}e^{-i\phi\ln S}d\phi$$

$$= \frac{e^{-r\tau}}{2\pi}\left[\int_{-\infty}^{\infty} f_2(S,v,\tau;-\phi)\int_{\ln K}^{\infty} e^y e^{i\phi y}dyd\phi\right.$$

$$\left. -K\int_{-\infty}^{\infty} f_2(S,v,\tau;-\phi)\int_{\ln K}^{\infty} e^{i\phi y}dyd\phi\right], \tag{5.65}$$

where we define $f_2(S,v,\tau;\phi) \equiv \exp\{B_2(\phi,\tau) + D_2(\phi,\tau)v + i\phi\ln S\}$.

We can evaluate the integrals in (5.65) using (5.71) and (5.72) from Appendix 5.B, provided that $f_2(S,v,\tau;-\phi)$ satisfies the appropriate assumptions. The first assumption we must verify is that $f_2(S,v,\tau;\phi - i)$ can be expressed as a function of ϕ. This assumption is satisfied, since

$$f_2(S,v,\tau;\phi - i) = Se^{(r-q)\tau}f_1(S,v,\tau;\phi),$$

where we define

$$f_1(S,v,\tau;\phi) \equiv \exp\{B_1(\phi,\tau) + D_1(\phi,\tau) + i\phi\ln S\},$$

with

$$B_1(\phi,\tau) = i\phi(r-q)\tau + \frac{\alpha}{\sigma^2}\left\{(\Theta_1 + \Omega_1)\tau - 2\ln\left(\frac{1 - Q_1 e^{\Omega_2\tau}}{1 - Q_1}\right)\right\}, \tag{5.66}$$

$$D_1(\phi,\tau) = \frac{(\Theta_1 + \Omega_1)}{\sigma^2}\left[\frac{1 - e^{\Omega_1\tau}}{1 - Q_1 e^{\Omega_1\tau}}\right], \tag{5.67}$$

and where $Q_1 = Q_1(\phi) \equiv (\Theta_1 + \Omega_1)/(\Theta_1 - \Omega_1)$, $\Theta_1 = \Theta_1(\phi) \equiv \Theta(i - \phi)$, and $\Omega_1 = \Omega_1(\phi) \equiv \Omega(i - \phi)$.

Furthermore, we can readily show that $f_j(S,v,\tau;-\phi) = \overline{f_j(S,v,\tau;\phi)}$ (where — denotes the complex conjugate) for $j = 1, 2$, and therefore all the assumptions required to carry out the calculations (5.71) and (5.72) are satisfied (these equations are defined in Appendix 5.B and are related to the calculation of some complex integrals). Thus (5.65) becomes

$$C_E(S,v,\tau)$$

$$= Se^{-q\tau}\left[\frac{1}{2}+\frac{1}{\pi}\int_0^\infty \mathrm{Re}\left(\frac{f_1(S,v,\tau;\phi)e^{-i\phi\ln K}}{i\phi}\right)d\phi\right]$$
$$- Ke^{-r\tau}\left[\frac{1}{2}+\frac{1}{\pi}\int_0^\infty \mathrm{Re}\left(\frac{f_2(S,v,\tau;\phi)e^{-i\phi\ln K}}{i\phi}\right)d\phi\right],$$

which is (5.34)–(5.35) in Proposition 5.8.

5.B Evaluation of Common Integral Terms in the Heston Model

In this appendix we consider complex integrals involving a complex function g that arise when pricing options under Heston's stochastic volatility dynamics. First we set up some notation that will be required in the calculations below. For the complex function g define the function h such that,

$$h(\phi) \equiv g(\phi - i).$$

It is not difficult to show that the function h satisfies,

$$h(-\phi) = \overline{h(\phi)}.$$

The integral terms that we need to consider are of the general form (see equation (5.65)), so we can write

$$J_1^H = \frac{1}{2\pi}\int_{-\infty}^\infty g(-\phi)\int_a^\infty e^y e^{i\phi y}dyd\phi, \tag{5.68}$$

and

$$J_2^H = \frac{1}{2\pi}\int_{-\infty}^\infty g(-\phi)\int_a^\infty e^{i\phi y}dyd\phi. \tag{5.69}$$

Beginning with J_1^H we have

$$J_1^H = \frac{1}{2\pi}\int_{-\infty}^\infty g(-\phi)\int_a^\infty e^{i(\phi-i)y}dyd\phi.$$

Changing the integration variable to $\psi = \phi - i$ gives

$$J_1^H = \frac{1}{2\pi}\int_{-\infty}^\infty g(-\psi - i)\int_a^\infty e^{i\psi y}dyd\psi.$$

Making the further change of variable $\psi = -\phi$ produces

$$J_1^H = \frac{1}{2\pi}\int_{-\infty}^\infty g(\phi - i)\int_a^\infty e^{-i\phi y}dyd\phi$$
$$= \frac{1}{2\pi}\int_{-\infty}^\infty g(\phi - i)\left[\lim_{b\to\infty}\frac{e^{-i\phi a} - e^{-i\phi b}}{i\phi}\right]d\phi.$$

Next we rewrite J_1^H as

$$J_1^H = \frac{1}{2\pi}\lim_{b\to\infty}\left[\int_0^\infty g(\phi - i)\left(\frac{e^{-i\phi a} - e^{-i\phi b}}{i\phi}\right)d\phi + \int_0^\infty g(-\phi - i)\left(\frac{e^{i\phi a} - e^{i\phi b}}{-i\phi}\right)d\phi\right]$$

$$= \frac{1}{2\pi}\int_0^\infty \frac{g(\phi - i)e^{-i\phi a} - g(-\phi - i)e^{i\phi a}}{i\phi}d\phi - \frac{1}{2\pi}\lim_{b\to\infty}\left[\int_0^\infty \frac{g(\phi - i)e^{-i\phi b} - g(-\phi - i)e^{i\phi b}}{i\phi}d\phi\right]. \quad (5.70)$$

Referring to Shephard (1991), we can evaluate the integral terms in (5.70) by way of the relation[9]

$$F(x) = \frac{1}{2} - \frac{1}{2\pi}\int_0^\infty \frac{f(\phi)e^{-i\phi x} - f(-\phi)e^{i\phi x}}{i\phi}d\phi,$$

where $F(x)$ denotes a cumulative density function. Defining $h(\phi) \equiv g(\phi - i)$, we can show that,

$$\lim_{b\to\infty}\frac{1}{2\pi}\int_0^\infty \frac{g(\phi - i)e^{-i\phi b} - g(-\phi - i)e^{i\phi b}}{i\phi}d\phi = \lim_{b\to\infty}\frac{1}{2\pi}\int_0^\infty \frac{h(\phi)e^{-i\phi b} - h(-\phi)e^{i\phi b}}{i\phi}d\phi = \lim_{b\to\infty}\left(\frac{1}{2} - F(b)\right) = -\frac{1}{2}.$$

Hence equation (5.70) becomes

$$J_1^H = \frac{1}{2} + \frac{1}{2\pi}\int_0^\infty \frac{h(\phi)e^{-i\phi a} - h(-\phi)e^{i\phi a}}{i\phi}d\phi.$$

Finally, given that $h(\phi)$ and $h(-\phi)$ are complex conjugates, that is, $h(-\phi) = \overline{h(\phi)}$, it follows that,

$$\frac{h(-\phi)e^{i\phi a}}{-i\phi} = \overline{\left(\frac{h(\phi)e^{-i\phi a}}{i\phi}\right)},$$

and thus we find that

$$J_1^H = \frac{1}{2} + \frac{1}{\pi}\int_0^\infty \mathrm{Re}\left(\frac{h(\phi)e^{-i\phi a}}{i\phi}\right)d\phi. \quad (5.71)$$

[9]The function F is defined by $F(x) = \int_{-\infty}^{x} f(u)du$.

For J_2^H, we make the change of variable let $\psi = -\phi$ to produce

$$J_2^H = \frac{1}{2\pi}\int_{-\infty}^{\infty} g(\psi)\int_a^{\infty} e^{-i\psi y}dyd\psi,$$

and similarly we can show that

$$J_2^H = \frac{1}{2} + \frac{1}{\pi}\int_0^{\infty} \text{Re}\left(\frac{g(\phi)e^{-i\phi a}}{i\phi}\right)d\phi, \tag{5.72}$$

where we assume that $g(-\phi) = \overline{g(\phi)}$.

5.C Calculation of the Deltas

First we calculate the delta expression under the free boundary approximation since this is required in the application of Newton's method to solve iteratively equations, this is essentially what equations (5.54) and (5.55) do. We recall that the expression for the American option price under the free boundary approximation is given in Proposition 5.12. With regard to the $\bar{f}_j$ functions defined in equation (5.46) we note that

$$\bar{f}_2(S, v, \tau; \phi - i, \phi a_1(\xi)) = Se^{(r-q)\tau}\bar{f}_1(S, v, \tau; \phi, -\phi a_1(\xi)). \tag{5.73}$$

(i) **American Call Delta — The free boundary approximation**

The delta for C_A is given by

$$\begin{aligned}
\frac{\partial C_A}{\partial S} &= Se^{-q\tau}\frac{\partial}{\partial S}\bar{P}_1^A(S, v, \tau; K, 0) + e^{-q\tau}\bar{P}_1^A(S, v, \tau; K, 0) \\
&\quad -Ke^{-r\tau}\frac{\partial}{\partial S}\bar{P}_2^A(S, v, \tau; K, 0) \\
&\quad + \int_0^{\tau} qe^{-q(\tau-\xi)}\left[S\frac{\partial}{\partial S}\bar{P}_1^A(S, v, \tau - \xi; e^{a_0(\xi)}, e^{a_1(\xi)})\right. \\
&\quad \left. + \bar{P}_1^A(S, v, \tau - \xi; e^{a_0(\xi)}, e^{a_1(\xi)})\right] d\xi \\
&\quad - \int_0^{\tau} rKe^{-r(\tau-\xi)}\frac{\partial}{\partial S}\bar{P}_2^A(S, v, \tau - \xi; e^{a_0(\xi)}, e^{a_1(\xi)})d\xi.
\end{aligned}$$

Consider

$$\frac{\partial}{\partial S}\bar{P}_j^A(S, v, \tau; e^{a_0}, e^{a_1}) = \frac{1}{\pi}\int_0^{\infty} \text{Re}\left(\frac{e^{-i\phi a_0}}{i\phi}\frac{\partial}{\partial S}\bar{f}_j(S, v, \tau; \phi_1, -a_1\phi)\right)d\phi,$$

where

$$
\begin{aligned}
\frac{\partial}{\partial S}\bar{f}_j(S,v,\tau;\phi,\psi) &= \frac{\partial}{\partial S}(\exp\{\bar{B}_j(\phi,\psi,\tau)+\bar{B}_j(\phi,\psi,\tau)v\}S^{i\phi}) \\
&= i\phi S^{i\phi-1}\exp\{\bar{B}_j(\phi,\psi,\tau)+\bar{B}_j(\phi,\psi,\tau)v\} \\
&= i\phi\exp\{\bar{B}_j(\phi,\psi,\tau)+\bar{B}_j(\phi,\psi,\tau)v+\ln S^{i(\phi+i)}\} \\
&= i\phi\exp\{\bar{B}_j(\phi,\psi,\tau)+\bar{B}_j(\phi,\psi,\tau)v+(i\phi-1)\ln S\} \\
&= \frac{i\phi}{S}\bar{f}_j(S,v,\tau;\phi,\psi).
\end{aligned}
$$

Hence

$$
\begin{aligned}
\frac{\partial C_A}{\partial S} &= Se^{-q\tau}\left(\frac{1}{\pi}\int_0^\infty \mathrm{Re}\left(\frac{e^{-i\phi\ln K}}{i\phi}\frac{i\phi}{S}\bar{f}_1(S,v,\tau;\phi,0)\right)d\phi\right) \\
&\quad -Ke^{-r\tau}\left(\frac{1}{\pi}\int_0^\infty \mathrm{Re}\left(\frac{e^{-i\phi\ln K}}{i\phi}\frac{i\phi}{S}\bar{f}_2(S,v,\tau;\phi,0)\right)d\phi\right) \\
&\quad +e^{-q\tau}\bar{P}_1^A(S,v,\tau;K,0) \\
&\quad +\int_0^\tau qe^{-q(\tau-\xi)}\bar{P}_1^A(S,v,\tau-\xi,e^{a_0(\xi)},e^{a_1(\xi)})d\xi \\
&\quad +\int_0^\tau\Bigg\{qSe^{-q(\tau-\xi)} \\
&\quad \times\left(\frac{1}{\pi}\int_0^\infty \mathrm{Re}\left(\frac{e^{-i\phi a_0(\xi)}}{i\phi}\frac{i\phi}{S}\bar{f}_1(S,v,\tau-\xi;\phi,-a_1(\xi)\phi)\right)d\phi\right) \\
&\quad -rKe^{-r(\tau-\xi)}\frac{1}{\pi}\int_0^\infty \mathrm{Re} \\
&\quad \times\left(\frac{e^{-i\phi a_0(\xi)}}{i\phi}\frac{i\phi}{S}\bar{f}_2(S,v,\tau-\xi;\phi,-a_1(\xi)\phi)\right)d\phi\Bigg\}d\xi.
\end{aligned}
$$

Using equation (5.73) this can be re–expressed as

$$
\begin{aligned}
\frac{\partial C_A}{\partial S} &= e^{-q\tau}\left(\frac{1}{\pi}\int_0^\infty \mathrm{Re}\left(e^{-i\phi\ln K}\frac{e^{(q-r)\tau}}{S}\bar{f}_2(S,v,\tau;\phi-i,0)\right)d\phi\right) \\
&\quad -\frac{Ke^{-r\tau}}{S}\left(\frac{1}{\pi}\int_0^\infty \mathrm{Re}\left(e^{-i\phi\ln K}\bar{f}_2(S,v,\tau;\phi,0)\right)d\phi\right) \\
&\quad +e^{-q\tau}\bar{P}_1^A(S,v,\tau;K,0)
\end{aligned}
$$

$$+\int_0^\tau qe^{-q(\tau-\xi)}\bar{P}_1^A(S,v,\tau-\xi;e^{a_0(\xi)},e^{a_1(\xi)})d\xi$$

$$+\int_0^\tau \left\{ qe^{-q(\tau-\xi)} \left(\frac{1}{\pi}\int_0^\infty \mathrm{Re}\left(e^{-i\phi a_0(\xi)}\frac{e^{(q-r)(\tau-\xi)}}{S}\right.\right.\right.$$

$$\left.\left.\times \bar{f}_2(S,v,\tau-\xi;\phi-i,a_1(\xi)\phi)\right)d\phi\right) - \frac{rKe^{-r(\tau-\xi)}}{S}$$

$$\left.\times\left(\frac{1}{\pi}\int_0^\infty \mathrm{Re}(e^{-i\phi a_0(\xi)}\bar{f}_2(S,v,\tau-\xi;\phi,-a_1(\xi)\phi))d\phi\right)\right\}d\xi.$$

Setting $\phi - i = \psi$ in the expression for $\bar{f}_j$ the last equation becomes

$$\frac{\partial C_A}{\partial S} = \frac{e^{-r\tau}}{S}\left(\frac{1}{\pi}\int_0^\infty \mathrm{Re}\left(e^{-i(\psi+i)\ln K}\bar{f}_2(S,v,\tau;\psi,0)\right)d\psi\right)$$

$$-\frac{Ke^{-r\tau}}{S}\left(\frac{1}{\pi}\int_0^\infty \mathrm{Re}\left(e^{-i\phi\ln K}\bar{f}_2(S,v,\tau;\phi,0)\right)d\phi\right)$$

$$+e^{-q\tau}\bar{P}_1^A(S,v,\tau;K,0)$$

$$+\int_0^\tau qe^{-q(\tau-\xi)}\bar{P}_1^A(S,v,\tau-\xi;e^{a_0(\xi)},e^{a_1(\xi)})d\xi$$

$$+\int_0^\tau \left\{\frac{qe^{-r(\tau-\xi)}}{S}\left(\frac{1}{\pi}\int_0^\infty \mathrm{Re}(e^{-i(\psi+i)(a_0(\xi))}\right.\right.$$

$$\left.\times \bar{f}_2(S,v,\tau-\xi;\psi,-a_1(\xi)(\psi+i)))d\phi\right) - \frac{rKe^{-r(\tau-\xi)}}{S}$$

$$\left.\times\left(\frac{1}{\pi}\int_0^\infty \mathrm{Re}\left(e^{-i\phi a_0(\xi)}\bar{f}_2(S,v,\tau-\xi;\phi_1-a_1(\xi))\phi\right)d\phi\right)\right\}d\xi$$

$$\frac{\partial C_A}{\partial S} = e^{-q\tau}\bar{P}_1^A(S,v,\tau;K,0) + \int_0^\tau qe^{-q(\tau-\xi)}\bar{P}_1^A(S,v,\tau-\xi;e^{a_0(\xi)},e^{a_1(\xi)})d\xi$$

$$+\int_0^\tau \frac{e^{-r(\tau-\xi)}}{S}\left(\frac{qe^{a_0(\xi)}}{\pi}\int_0^\infty \mathrm{Re}(e^{-i\phi a_0(\xi)}\right.$$

$$\times \bar{f}_2(S,v,\tau-\xi;\phi,-a_1(\xi)(\phi+i)))d\phi$$

$$\left.-\frac{rK}{\pi}\int_0^\infty \mathrm{Re}(e^{-i\phi a_0(\xi)}\bar{f}_2(S,v,\tau-\xi;\phi,-a_1(\xi)\phi))d\phi\right)d\xi.$$

Note that unlike the constant volatility case, we cannot simplify the last integral term any further.

(ii) **Newton's Method applied to equations (5.54) and (5.55)**

Let

$$F(a_1(\tau)) = \frac{1}{v}(\ln[C_A(e^{a_0(\tau)+va_1(\tau)}, v, \tau) + K] - a_0(\tau)) - a_1(\tau),$$

from which we calculate

$$\frac{\partial F}{\partial a_1(\tau)} = \frac{1}{v}\left(\frac{1}{C_A(e^{a_0(\tau)+va_1(\tau)}, v, \tau) + K}\frac{\partial}{\partial a_1(\tau)}C_A(e^{a_0(\tau)+va_1(\tau)}, v, \tau)\right) - 1, \tag{5.74}$$

where

$$\frac{\partial}{\partial a_1(\tau)}\{C_A(e^{a_0(\tau)+va_1(\tau)}, v, \tau)\} = \left.\frac{\partial C_A(S, v, \tau)}{\partial S}\right|_{S=e^{a_0(\tau)+va_1(\tau)}} ve^{a_0(\tau)+va_1(\tau)}.$$

Next let

$$G(a_0(\tau)) = \ln\left[C_A(e^{a_0(\tau)+va_1(\tau)}, v, \tau) + K\right] - va_1(\tau) - a_0(\tau),$$

so that

$$\frac{\partial G}{\partial a_0(\tau)} = \frac{1}{C_A(e^{a_0(\tau)+va_1(\tau)}, v, \tau) + K}\frac{\partial}{\partial a_0(\tau)}C_A(e^{a_0(\tau)+va_1(\tau)}, v, \tau) - 1, \tag{5.75}$$

where

$$\frac{\partial}{\partial a_0(\tau)}\{C_A(e^{a_0(\tau)+va_1(\tau)}, v, \tau)\} = \left.\frac{\partial C_A(S, v, \tau)}{\partial S}\right|_{S=e^{a_0(\tau)+va_1(\tau)}} \cdot e^{a_0(\tau)+va_1(\tau)}.$$

The derivatives (5.74) and (5.75) are required for the implementation of Newton's method.

(iii) **American Call Delta (The general case)**

In the general case, the expression for the American call price is given in Proposition 5.10.

Differentiating with respect to S we obtain

$$\frac{\partial C_A}{\partial S} = Se^{-q\tau}\frac{\partial P_1^H}{\partial S} + e^{-q\tau}P_1^H(S, v, \tau; K) - Ke^{-r\tau}\frac{\partial P_2^H}{\partial S}$$

$$+\int_0^\tau \int_0^\infty qe^{-q(\tau-\xi)} \left[S\frac{\partial P_1^A}{\partial S} + P_1^A(S, v, \tau - \xi; w, a(w, \xi)) \right] dwd\xi$$
$$-\int_0^\tau \int_0^\infty rKe^{-r(\tau-\xi)} \frac{\partial P_2^A}{\partial S} dwd\xi.$$

Firstly, consider

$$\frac{\partial P_1^H}{\partial S} = \frac{\partial}{\partial S}\left\{ \frac{1}{2} + \frac{1}{\pi}\int_0^\infty \mathrm{Re}\left(\frac{e^{-i\phi \ln K}}{i\phi} \exp\{B_1(\phi, \tau) + D_1(\phi, \tau)v \right.\right.$$
$$\left.\left. + i\phi \ln S\} \right) d\phi \right\}$$
$$= \frac{1}{\pi}\int_0^\infty \mathrm{Re}\left(\frac{e^{-i\phi \ln K}}{S} f_1(S, v, \tau; \phi) \right) d\phi.$$

Since

$$f_2(S, v, \tau; \phi - i) = Se^{(r-q)\tau} f_1(S, v, \tau; \phi),$$

we have

$$\frac{\partial P_1^H}{\partial S} = \frac{1}{\pi}\int_0^\infty \mathrm{Re}\left(\frac{e^{-i\phi \ln K}}{S^2} e^{(q-r)\tau} f_2(S, v, \tau; \phi - i) \right) d\phi.$$

Let $\psi = \phi - i$, then

$$\frac{\partial P_1^H}{\partial S} = \frac{1}{\pi}\int_0^\infty \mathrm{Re}\left(e^{-i\phi \ln K} f_2(S, v, \tau; \phi) \frac{e^{(q-r)\tau}}{S^2} e^{\ln K} \right) d\phi$$
$$= \frac{1}{\pi}\int_0^\infty \frac{K}{S^2} e^{(q-r)\tau} \mathrm{Re}(e^{-i\phi \ln K} f_2(S, v, \tau; \phi)) d\phi.$$

Similarly we calculate

$$\frac{\partial P_2^H}{\partial S} = \frac{1}{\pi}\int_0^\infty \mathrm{Re}\left(\frac{e^{-i\phi \ln K}}{S} f_2(S, v, \tau; \phi) \right) d\phi.$$

In the early exercise premium term

$$\frac{\partial P_j^A}{\partial S} = \frac{1}{\pi}\int_0^\infty \mathrm{Re}\left(\frac{e^{-i\phi \ln a(w,\xi)}}{S} g_j(S, v, \tau - \xi; \phi, w) \right) d\phi, \quad \text{for } j = 1, 2.$$

Thus the delta for the American call is given by

$$\begin{aligned}\frac{\partial C_A}{\partial S} &= e^{-q\tau}P_1^H(S,v,\tau;K)\\ &\quad+\int_0^\tau\int_0^\infty qe^{-q(\tau-\xi)}P_1^A(S,v,\tau-\xi;w,a(w,\xi))dwd\xi\\ &\quad+\int_0^\tau\int_0^\infty qe^{-q(\tau-\xi)}\\ &\quad\times\left[\frac{1}{\pi}\int_0^\infty \mathrm{Re}(e^{-i\phi\ln a(w,\xi)}g_1(S,v,\tau-\xi;\phi,w))d\phi\right]dwd\xi\\ &\quad-\int_0^\tau\int_0^\infty \frac{rK}{S}e^{-r(\tau-\xi)}\\ &\quad\times\left[\frac{1}{\pi}\int_0^\infty \mathrm{Re}(e^{-i\phi\ln a(w,\xi)}g_2(S,v,\tau-\xi;\phi,w))d\phi\right]dwd\xi.\end{aligned}$$

5.D Proof of Proposition 5.12

In deriving (5.44), we need only attend to the early exercise premium, $C_P(S,v,\tau)$, and how it is affected by the assumption for $a(v,\tau)$ in (5.43). Using the Duhamel's principle and (5.43), equation (5.14) for $C_P(S,v,\tau)$ becomes

$$\begin{aligned}&C_P(S,v,\tau)\\ &\quad=\int_0^\tau e^{-r(\tau-\xi)}\int_0^\infty\int_{a_0(\xi)+a_1(\xi)w}^\infty (qe^y-rK)G(y,w,\tau-\xi;S,v)dydwd\xi.\end{aligned}$$

Making the change of integration variable $z=y-a_1(\xi)w$, along with a change in the order of integration, we have

$$\begin{aligned}C_P(S,v,\tau) &= \int_0^\tau e^{-r(\tau-\xi)}\int_{a_0(\xi)}^\infty\int_0^\infty (qe^{z+a_1(\xi)w}-rK)G(z+a_1(\xi),w,\tau\\ &\qquad -\xi;S,v)dwdzd\xi\\ &= \int_0^\tau e^{-r(\tau-\xi)}\int_{a_0(\xi)}^\infty [qe^z\bar{J}_1(z,\xi)-rK\bar{J}_2(z,\xi)]dzd\xi, \qquad (5.76)\end{aligned}$$

where we set

$$\bar{J}_1(z,\xi)\equiv\int_0^\infty e^{a_1(\xi)w}G(z+a_1(\xi),w,\tau-\xi;S,v)dw,$$

and

$$\bar{J}_2(z,\xi)\equiv\int_0^\infty G(z+a_1(\xi),w,\tau-\xi;S,v)dw.$$

We begin by considering $\bar{J}_2$. Using the definition of G from (5.32) in conjunction with a change of variables given by (4.77) and (4.80) in Chapter 4, we find that

$$\begin{aligned}
&\bar{J}_2(z,\xi) \\
&\quad = \frac{1}{2\pi}\int_{-\infty}^{\infty} e^{i\phi(z-\ln S)} e^{-i\phi(r-q)(\tau-\xi)} \exp\left\{\left[\frac{\alpha(\Theta-\Omega)}{\sigma^2}\right](\tau-\xi)\right\} \\
&\quad\quad \times \left[\int_0^{\infty} e^{-A-y} \exp\left\{-\left(\frac{\Theta-\sigma^2 i\phi a_1(\xi)-\Omega}{\sigma^2}\right)\frac{\sigma^2(1-e^{-\Omega(\tau-\xi)})}{2\Omega}A\right\}\right. \\
&\quad\quad \times \left.\left(\frac{A}{y}\right)^{\frac{\alpha}{\sigma^2}-\frac{1}{2}} \exp\left\{\frac{(\Theta-\Omega)(e^{\Omega(\tau-\xi)}-1)}{2\Omega}y\right\} I_{\frac{2\alpha}{\sigma^2}-1}(2\sqrt{Ay})dA\right] d\phi.
\end{aligned} \tag{5.77}$$

Referring to the results in Appendix 5.A for the European call, we note that (5.77) is very similar in form to (5.60),[10] and this allows us to make use of the same procedure that led from (5.60) to (5.61) to produce

$$\begin{aligned}
\bar{J}_2(z,\xi) &= \frac{1}{2\pi}\int_{-\infty}^{\infty} e^{i\phi(z-\ln S)} e^{-i\phi(r-q)(\tau-\xi)} e^{\frac{(\Theta-\Omega)}{\sigma^2}(\alpha(\tau-\xi)+v)} \\
&\quad \times \left(\frac{2\Omega}{(\Theta-\sigma^2 i\phi a_1(\xi)-\Omega)(1-e^{-\Omega(\tau-\xi)})+2\Omega}\right)^{\frac{2\alpha}{\sigma^2}} \\
&\quad \times \exp\left\{\frac{-2\Omega v}{\sigma^2(e^{\Omega(\tau-\xi)}-1)}\right\} \\
&\quad \times \exp\left\{\frac{2\Omega}{(\Theta-\sigma^2 i\phi a_1(\xi)-\Omega)(1-e^{-\Omega(\tau-\xi)})+2\Omega}\right. \\
&\quad \times \left.\frac{2\Omega v}{\sigma^2(e^{\Omega(\tau-\xi)}-1)}\right\} d\phi.
\end{aligned} \tag{5.78}$$

The last step is to express (5.78) in a form that is analogous to the European case. Specifically, we write $\bar{J}_2$ as

$$\bar{J}_2(z,\xi) = \frac{1}{2\pi}\int_{-\infty}^{\infty} e^{i\phi z} \bar{f}_2(S,v,\tau-\xi;-\phi,\phi a_1(\xi))d\phi, \tag{5.79}$$

where we define

$$\bar{f}_2(S,v,\tau;\phi,\psi) \equiv \exp\{\bar{B}_2(\phi,\psi,\tau)+\bar{D}_2(\phi,\psi,\tau)v+i\phi\ln S\},$$

[10] This comparison is more readily made by taking the inverse Fourier transform of (5.60).

with

$$\bar{B}_2(\phi,\psi,\tau) = i\phi(r-q)\tau + \frac{\alpha}{\sigma^2}\left\{(\Theta_2+\Omega_2)\tau - 2\ln\left(\frac{1-\bar{Q}_2e^{\Omega_2\tau}}{1-\bar{Q}_2}\right)\right\},$$

$$\bar{D}_2(\phi,\psi,\tau) = i\psi + \frac{(\Theta_2-\sigma^2 i\psi+\Omega_2)}{\sigma^2}\left(\frac{1-e^{\Omega_2\tau}}{1-\bar{Q}_2e^{\Omega_2\tau}}\right),$$

and $\bar{Q}_2 = (\Theta_2-\sigma^2 i\psi+\Omega_2)/(\Theta_2-\sigma^2 i\psi-\Omega_2)$. The terms Θ_2 and Ω_2 are given by $\Theta_2 = \Theta_2(\phi) \equiv \Theta(-\phi)$ and $\Omega_2 = \Omega_2(\phi) \equiv \Omega(-\phi)$.

Next we repeat this process for the term $\bar{J}_1$. As for $\bar{J}_2$, we use (5.32), in conjunction with (4.77) and (4.80), to give

$$\begin{aligned}\bar{J}_1(z,\xi) = \frac{1}{2\pi}\int_{-\infty}^{\infty} & e^{i\phi(z-\ln S)}e^{-i\phi(r-q)(\tau-\xi)}\exp\left\{\left[\frac{\alpha(\Theta-\Omega)}{\sigma^2}\right](\tau-\xi)\right\}\\ &\times\left[\int_0^\infty e^{-A-y}\exp\left\{-\left(\frac{\Theta-\sigma^2 i(\phi-i)a_1(\xi)-\Omega}{\sigma^2}\right)\right.\right.\\ &\times\left.\frac{\sigma^2(1-e^{-\Omega(\tau-\xi)})}{2\Omega}A\right\}\\ &\times\left.\left(\frac{A}{y}\right)^{\frac{\alpha}{\sigma^2}-\frac{1}{2}}\exp\left\{\frac{(\Theta-\Omega)(e^{\Omega(\tau-\xi)}-1)}{2\Omega}y\right\}I_{\frac{2\alpha}{\sigma^2}-1}(2\sqrt{Ay})dA\right]d\phi.\end{aligned} \tag{5.80}$$

By noting that (5.77) and (5.80) only differ by the fact that the coefficient of $a_1(\xi)$ in (5.80) is $(\phi-i)$ where is ϕ in (5.77), we can readily write (5.80) as

$$\bar{J}_1(z,\xi) = \frac{1}{2\pi}\int_{-\infty}^{\infty} e^{i\phi z}\bar{f}_2(S,v,\tau-\xi;-\phi,(\phi-i)a_1(\xi))d\phi. \tag{5.81}$$

We now substitute (5.79) and (5.81) into (5.76), and find that

$$\begin{aligned}C_P(S,v,\tau) = \int_0^\tau & \frac{e^{-r(\tau-\xi)}}{2\pi}\\ &\times\left[\int_{-\infty}^{\infty}\bar{f}_2(S,v,(\tau-\xi);-\phi,(\phi-i)a_1(\xi))\int_{a_0(\xi)}^{\infty}e^z e^{i\phi z}dzd\phi\right.\\ &\left.-K\int_{-\infty}^{\infty}\bar{f}_2(S,v,(\tau-\xi);-\phi,\phi a_1(\xi))\int_{a_0(\xi)}^{\infty}e^{i\phi z}dzd\phi\right]d\xi.\end{aligned} \tag{5.82}$$

As in the European call case, we can evaluate the integrals with respect to z and ϕ in (5.82) by use of (5.71) and (5.72) from Appendix 5.B, provided that $\bar{f}_2(S,v,\tau;-\phi,\psi)$ satisfies the appropriate assumptions for the given values of ψ. The first assumption requires that we must verify that $\bar{f}_2(S, v, \tau; \phi - i, -\phi a_1(\xi))$ can be expressed as a function of ϕ. This assumption is satisfied, since

$$\bar{f}_2(S, v, \tau; \phi - i, -\phi a_1(\xi)) = Se^{(r-q)\tau} \bar{f}_1(S, v, \tau; \phi, -\phi a_1(\xi)),$$

where we define

$$\bar{f}_1(S, v, \tau; \phi, \psi) \equiv \exp\{\bar{B}_1(\phi, \psi, \tau) + \bar{D}_1(\phi, \psi, \tau) + i\phi \ln S\},$$

with

$$\bar{B}_1(\phi, \psi, \tau) = i\phi(r-q)\tau + \frac{\alpha}{\sigma^2}\left\{(\Theta_1 + \Omega_1)\tau - 2\ln\left(\frac{1 - \bar{Q}_1 e^{\Omega_1 \tau}}{1 - \bar{Q}_1}\right)\right\},$$

$$\bar{D}_1(\phi, \psi, \tau) = i\psi + \frac{(\Theta_1 - \sigma^2 i\psi + \Omega_1)}{\sigma^2}\left(\frac{1 - e^{\Omega_1 \tau}}{1 - \bar{Q}_1 e^{\Omega_1 \tau}}\right),$$

and $\bar{Q}_1 = (\Theta_1 - \sigma^2 i\psi + \Omega_1)/(\Theta_1 - \sigma^2 i\psi - \Omega_1)$, $\Theta_1 = \Theta_1(\phi) \equiv \Theta(i - \phi)$, and $\Omega_1 = \Omega_1(\phi) \equiv \Omega(i - \phi)$.

Furthermore, we can readily show that

$$f_j(S, v, \tau; -\phi, \phi a_1(\xi)) = \overline{\bar{f}_j(S, v, \tau; \phi, -\phi a_1(\xi))} \quad \text{for } j = 1, 2$$

and therefore all the assumptions required by (5.71) and (5.72) are satisfied. Thus (5.82) becomes

$$\begin{aligned} C_P(S, v, \tau) = \int_0^\tau \Bigg(& Se^{-q(\tau-\xi)}\left[\frac{1}{2} + \frac{1}{\pi}\int_0^\infty \text{Re}\right. \\ & \times \left.\left(\frac{\bar{f}_1(S, v, \tau - \xi; \phi, -\phi a_1(\xi))e^{-i\phi a_0(\xi)}}{i\phi}\right) d\phi\right] \\ & - Ke^{-r(\tau-\xi)}\left[\frac{1}{2} + \frac{1}{\pi}\int_0^\infty \text{Re}\right. \\ & \left.\left(\frac{\bar{f}_2(S, v, \tau - \xi; \phi, -\phi a_1(\xi))e^{-i\phi a_0(\xi)}}{i\phi}\right) d\phi\right]\Bigg) d\xi, \end{aligned}$$

which is the early exercise premium component of (5.44) in Proposition 5.12.

5.E Moments for the Heston Model

We recall that the dynamics for the Heston model are given by (5.1) and (5.2). The first two moments for v are (see Dufresne (2001))

$$\mathbb{E}_0[v_t] = \Theta + (v_0 - \Theta)e^{-\kappa t},$$

$$\mathbb{E}_0[v_t^2] = \frac{(2\kappa\Theta + \sigma^2)\Theta}{2\kappa} + \frac{(2\kappa\Theta + \sigma^2)}{\kappa}(v_0 - \Theta)e^{-\kappa t} + \left[\frac{(2\kappa\Theta + \sigma^2)\Theta}{2\kappa} - \frac{(2\kappa\Theta + \sigma^2)}{\kappa}v_0 + v_0^2\right]e^{-2\kappa t},$$

and hence the variance for v is

$$\text{Var}_0[v_t] = \frac{\Theta\sigma^2}{2\kappa} - \left(\frac{\theta\sigma^2}{\kappa} - \frac{\sigma^2}{\kappa}v_0\right)e^{-\kappa t} + \left(\frac{\theta\sigma^2}{2\kappa} - \frac{\sigma^2}{\kappa}v_0\right)e^{-2\kappa t}. \quad (5.83)$$

Using the characteristic function for $\ln S$ (given by Heston (1993)) we can also calculate the moments of $\ln S$.

The first moment for $\ln S_\tau$ (where $\tau = T - t$) is

$$\mathbb{E}_T[\ln S_\tau] = \frac{e^{-\beta\tau}}{2\beta^2}\{2e^{\beta\tau}\beta^2 \ln S_0 - \alpha[1 + e^{\beta\tau}(\beta\tau - 1)] + \beta[(e^{\beta\tau - 1})v_0 - 2e^{\beta\tau}\mu\beta\tau]\},$$

where $\alpha = \kappa\Theta$ and $\beta = \kappa + \lambda$.

The variance is

$$\text{Var}_T[\ln S_\tau] = \frac{e^{-2\beta\tau}}{8\beta^4}\{(\alpha - 2v_0\beta)\sigma^2 + 4e^{\beta\tau}f(v_0,\tau) + e^{2\beta\tau}g(v_0,\tau)\}$$

where

$$\begin{aligned} f(v_0,\tau) &= \alpha[\sigma^2 + \beta\sigma(\sigma\tau - 4\rho) + 2\beta^2(1 - \rho\sigma\tau)] \\ &\quad + v_0\beta^2[\sigma(2\rho - \sigma\tau) + 2\beta(\rho\sigma\tau - 1)], \\ g(v_0,\tau) &= 2v_0\beta[4\beta(\beta - \rho\sigma) + \sigma^2] \\ &\quad + \alpha[8\beta^3\tau - 5\sigma^2 + 2\beta\sigma(8\rho + \sigma\tau) - 8\beta^2(1 + \rho\sigma\tau)]. \end{aligned}$$

When comparing the Black-Scholes model

$$dS = \mu S dt + \sigma_{\text{BS}} S dZ$$

with the Heston model, we match the variances of $\ln S$ according to

$$\sigma_{\text{BS}}^2\tau = \text{Var}_T[\ln S_\tau].$$

5.F Method of Lines for the Heston PDE

There is no closed-form solution for American call prices, even when the volatility is constant. Thus in order to analyse the quality of the numerical integration method in Sec. 5.5 we need to consider other numerical solution methods. One such benchmark involves the use of finite difference methods, such as the operator splitting technique employed by Ikonen and

Toivanen (2004). Here we consider another approach using the method of lines technique. The method of lines has been applied to American options with constant volatility by Meyer and van der Hoek (1997), and generalised to the jump-diffusion setting by Meyer (1998). Here we present an extension of the approach by Meyer and van der Hoek (1997) to solve the two-factor free boundary value problem (5.3)–(5.8).

The purpose of the method of lines is to replace a PDE with an equivalent system of one-dimensional ODEs, whose solution can be readily found numerically. When volatility is constant, the system of ODEs is developed by discretising the time derivative. For the Heston PDE (5.3), we must also discretise the derivative terms involving the volatility v. We begin by setting $v_m = m\Delta v$, where $m = 0, 1, 2, \ldots, M$. Typically we will set the maximum volatility to be $v_M = 100\%$. Furthermore, we discretise the time to expiry according to $\tau_n = n\Delta\tau$, where $\tau_N = T$. We denote the option price along the volatility line v_m and time line τ_n by $C_A(S, v_m, \tau_n) = C_m^n(S)$, and define

$$V_A(S, v_m, \tau_n) \equiv \frac{\partial C_A(S, v_m, \tau_n)}{\partial S} = V_m^n(S). \tag{5.84}$$

We now select finite difference approximations for the deritvative terms with respect to v. For the second order term we use the standard central difference scheme

$$\frac{\partial^2 C_m^n}{\partial v^2} = \frac{C_{m+1}^n - 2C_m^n + C_{m-1}^n}{(\Delta v)^2}, \tag{5.85}$$

and similarly for the cross derivative term, we use the central difference approximation

$$\frac{\partial^2 C_m^n}{\partial S \partial v} = \frac{V_{m+1}^n - V_{m-1}^n}{2\Delta v}. \tag{5.86}$$

Since the coefficients of the second order derivative terms go to zero as $v \to 0$, we use an upwinding finite difference scheme for the first order derivative term, such that

$$\frac{\partial C_A}{\partial v} = \begin{cases} \frac{C_{m+1}^n - C_m^n}{\Delta v} & \text{if } v \leq \frac{\alpha}{\beta}, \\ \frac{C_m^n - C_{m-1}^n}{\Delta v} & \text{if } v > \frac{\alpha}{\beta}. \end{cases} \tag{5.87}$$

Since the second order derivative terms both vanish as $v \to 0$, upwinding helps to stabilise the finite difference scheme with respect to v.

Next we must select a discretisation for the time derivative. Initially we use a standard backward difference, given by

$$\frac{\partial C_m^n}{\partial \tau} = \frac{C_m^n - C_m^{n-1}}{\Delta \tau}. \tag{5.88}$$

This approximation is only first order accurate with respect to time. Meyer and van der Hoek (1997) demonstrate that the accuracy of the method of lines increases considerably by using a second order approximation for the time derivative, namely

$$\frac{\partial C_m^n}{\partial \tau} = \frac{3}{2}\frac{C_m^n - C_m^{n-1}}{\Delta \tau} + \frac{1}{2}\frac{C_m^{n-1} - C_m^{n-2}}{\Delta \tau}. \tag{5.89}$$

Thus we initiate the method of lines solution by using (5.88) for the first several time steps, and then switching to (5.89) for all subsequent time steps.

Applying (5.84)–(5.89) to the PDE (5.3), we must now solve a system of second order ODEs at each time step. For the first few time steps, the ODE for a given value of v is

$$\begin{aligned} &\frac{v_m S^2}{2}\frac{d^2 C_m^n}{dS^2} + \rho\sigma v_m S \frac{V_{m+1}^n - V_{m-1}^n}{2\Delta v} + \frac{\sigma^2 v_m}{2}\frac{C_{m+1}^n - 2C_m^n + C_{m-1}^n}{(\Delta v)^2} \\ &\quad + (r-q)S\frac{dC_m^n}{dS} + \frac{1}{2}(\alpha - \beta v + |\alpha - \beta v|)\frac{C_{m+1}^n - C_m^n}{\Delta v} \\ &\quad + \frac{1}{2}(\alpha - \beta v - |\alpha - \beta v|)\frac{C_m^n - C_{m-1}^n}{\Delta v} - rC_m^n - \frac{C_m^n - C_m^{n-1}}{\Delta \tau} = 0, \end{aligned} \tag{5.90}$$

and for all subsequent time steps the ODE is

$$\begin{aligned} &\frac{v_m S^2}{2}\frac{d^2 C_m^n}{dS^2} + \rho\sigma v_m S \frac{V_{m+1}^n - V_{m-1}^n}{2\Delta v} + \frac{\sigma^2 v_m}{2}\frac{C_{m+1}^n - 2C_m^n + C_{m-1}^n}{(\Delta v)^2} \\ &\quad + (r-q)S\frac{dC_m^n}{dS} + \frac{1}{2}(\alpha - \beta v + |\alpha - \beta v|)\frac{C_{m+1}^n - C_m^n}{\Delta v} \\ &\quad + \frac{1}{2}(\alpha - \beta v - |\alpha - \beta v|)\frac{C_m^n - C_{m-1}^n}{\Delta v} \\ &\quad - rC_m^n - \frac{3}{2}\frac{C_m^n - C_m^{n-1}}{\Delta \tau} + \frac{1}{2}\frac{C_m^{n-1} - C_m^{n-2}}{\Delta \tau} = 0. \end{aligned} \tag{5.91}$$

We requite two boundary conditions in the v direction, one at v_0 and the other at v_M. For the v_M boundary condition, we note that for large values of v the option price changes slowly, and thus we set $\partial C_M/\partial v = 0$. To handle the boundary condition at $v = 0$ we fit a quadratic polynomial through the option prices at v_1, v_2 and v_3, and then use this to extrapolate an approximation of the price at v_0. This provides us with a satisfactory estimate of the price along v_0 for the purpose of generating a stable solution for small values of v.

After taking the boundary conditions into consideration, at each time step n we must solve a system of $M - 1$ second order ODEs. We solve the ODEs for increasing values of v, using the latest available estimates for C^n_{m+1}, C^n_{m-1}, V^n_{m+1} and V^n_{m-1}. The initial estimates for C^n_m and V^n_m are simply C^{n-1}_m and V^{n-1}_m. Otherwise we use the latest estimates for C^n_m and V^n_m found during the current iteration through the volatility lines. We iterate until the price profile converges to a desired level of accuracy.

The generic first order form for (5.90) and (5.91) is

$$\frac{dC^n_m}{dS} = V^n_m, \tag{5.92}$$

$$\frac{dV^n_m}{dS} = A_m(S)C^n_m + B_m(S)V^n_m + g_m(S), \tag{5.93}$$

where $g_m(S)$ is a function of C^n_{m+1}, C^n_{m-1}, V^n_{m+1}, V^n_{m-1}, C^{n-1}_m and C^{n-2}_m.

The solution of (5.92) and (5.93) is related to the Riccati transformation

$$C^n_m(S) = R_m(S)V^n_m(S) + w^n_m(S) \tag{5.94}$$

where R and w are the solutions of

$$\frac{dR_m}{dS} = 1 - B_m(S)R_m(S) - A_m(S)R_m(S)^2, \quad R_m(0) = 0, \tag{5.95}$$

$$\frac{dw^n_m}{dS} = -A_m(S)R_m(S)w^n_m(S) - R_m(S)g^n_m(S), \quad w^n_m(0) = 0. \tag{5.96}$$

Equations (5.95) and (5.96) can be solved numerically. Once R and w are known, the free boundary, $a(v, \tau)$, can be be determined by finding the roots of (5.97) using boundary condition (5.52) and the first condition of (5.53), namely

$$a(v_m, \tau_n) - K = R_m(a(v_m, \tau_n)) + w^n_m(a(v_m, \tau_n)) \tag{5.97}$$

We solve for $a(v_m, \tau_m)$ using Newton's method. Once we have $a(v_m, \tau_m)$ we can find V^n_m by substituting (5.94) into (5.93) and integrating numerically the ODE. That is to say

$$\frac{dV_m^n}{dS} = A_m(S)[R_m(S)V_m^n(S) + w_m^n(S)] + B_m(S)V_m^n + g_m^n(S), \tag{5.98}$$

subject to the end-point condition

$$V_m^n(a_m^n) = 1.$$

The American option price is then determined by substituting R, w, V into the Riccati transformation (5.94).

A somewhat more detailed explanation of the method of lines is given in Sec. 7.4 where it is applied to the American call under jump-diffusion and stochastic volatility.

Chapter 6

Fourier Cosine Expansion Approach

The Fourier cosine expansion approach (COS) is developed by Fang and Oosterlee (2008) using the Cosine series expansions of the value function at the next time level and the density function. The resulting equation is called the COS formula, due to the use of Fourier cosine series expansions. Fourier series expansions and their convergence properties have been discussed in Fang and Oosterlee (2008), so we do not go into details here but refer the reader to this excellent paper.

6.1 Heston Model

For the underlying dynamics, we assume that the stochastic differential equation (SDE) for S is given by the square root variance process of Heston (1993), summarised by equations (6.1) and (6.2). Thus the dynamics for S and under the so-called risk neutral measure $\mathbb{Q}$ are governed by the SDE system

$$dS = (r-q)Sdt + \sqrt{v}Sd\tilde{Z}_1, \tag{6.1}$$

$$dv = \kappa_v(\theta - v)dt + \sigma_v\sqrt{v}d\tilde{Z}_2, \tag{6.2}$$

where $\tilde{Z}_1$ and $\tilde{Z}_2$ are Wiener processes under the risk neutral measure.

We need to focus on the near-singular behavior of the variance process near the origin. The square-root process precludes negative values for v_t, and if v_t reaches zero, it can subsequently become positive. The Feller condition $2\kappa_v\theta \geq \sigma_v^2$, guarantees that v stays positive, otherwise it may reach zero. As indicated in Cox *et al.* (1985), with

$$\eta = 2\kappa_v\theta/\sigma_v^2 - 1, \quad \text{and} \quad \zeta = 2\kappa_v/((1-e^{-\kappa_v(t-s)})\sigma_v^2), \tag{6.3}$$

the process $2\zeta v_t \sim \chi^2(q, 2\zeta v_s e^{-\kappa_v(t-s)})$, for $0 < s < t$, is governed by the noncentral chi-square distribution with degree η and noncentrality parameter $2\zeta v_s e^{-\kappa_v(t-s)}$. Therefore, the probability density function of v_t given

v_s reads as

$$p_v(v_t|v_s) = \zeta e^{-\zeta(v_s e^{-\kappa_v(t-s)}+v_t)} \left(\frac{v_t}{v_s e^{-\kappa_v(t-s)}}\right)^{\frac{\eta}{2}} \times I_\eta(2\zeta e^{-\frac{1}{2}\kappa_v(t-s)}\sqrt{v_s v_t}), \tag{6.4}$$

where $I_\eta(\cdot)$ is the modified Bessel function of the first kind with order η (see equation (4.24)).

The Feller condition is thus equivalent to "$\eta \geq 0$". This is difficult to satisfy in practice, in which case the cumulative distribution of the variance shows a near-singular behavior near the origin, or, in other words, the left tail of the variance density grows extremely fast in value. Such a behavior in the left tail may easily give rise to significant errors, especially for integration-based option pricing methods, for which the integration range needs to be truncated. In Fang and Oosterlee (2011), the authors discuss in more detail how the parameter values, especially η will affect left tail of the distribution of v_t based on which the truncation range will be determined.

6.1.1 *Transformation to the log-variance process*

In order to propose a solution for the problem of the left-side tail, we transform the problem from the variance domain to the log-variance domain. By this change of variable, the density of the log-variance process, based on (6.4), becomes

$$p_{\ln(v)}(u_t|u_s) = \zeta e^{-\zeta(e^{u_s} e^{-\kappa_v(t-s)}+e^{u_t})} \left(\frac{e^{u_t}}{e^{u_s} e^{-\kappa_v(t-s)}}\right)^{\frac{\eta}{2}} I_\eta(2\zeta e^{-\frac{1}{2}\kappa_v(t-s)}\sqrt{e^{u_s} e^{u_t}}), \tag{6.5}$$

where $u_s = \ln(v_s)$ and $p_{\ln v}(u_t|u_s)$ denotes the probability density of the log-variance at a future time, given the information at the current time.

With the change of variables, a term e^{u_t} appears, which, for $\eta \in [-1, 0]$, compensates the $(\cdot)^{\frac{\eta}{2}}$ term, so that it converges towards zero as $u_t \to -\infty$. Hence, the left tails of the densities no longer increase significantly in value. Instead, these tails decay to zero rapidly as $u_t \to -\infty$, although the decay rate decreases as η approaches -1.

When valuing American options or path-dependent options, we need to know the joint distribution of the log-stock and log-variance processes at a future time, given the information at current time, that is, $p_{x,\ln(v)}(x_t, u_t|x_s, u_s)$ with $0 < s < t$. However an analytic formula for this distribution does not exist, but we can deduce the relevant information from the Fourier domain. The Heston SDEs indicate that the variance at a future time is independent of the log-stock value at the current time, so that $p_v(v_t|v_s, x_s) = p_v(v_t|v_s)$. Hence, we have

$$p_{x,v}(x_t, v_t|x_s, v_s) = p_{x|v}(x_t|v_t, x_s, v_s)p_v(v_t|v_s), \tag{6.6}$$

where $p_{x,v}$ denotes the joint probability density of the log-stock and the variance processes at a future time point t, given that the information is known at the current time s; $p_{x|v}$ denotes the probability density of the log-stock process at a future time point t, given the variance value at current time s. Similarly, we have,

$$p_{x,\ln(v)}(x_t, u_t|x_s, u_s) = p_{x|\ln(v)}(x_t|u_t, x_s, u_s)p_{\ln(v)}(u_t|u_s), \tag{6.7}$$

where $p_{x,\ln(v)}$ denotes the probability density of the log-stock at a future time spot price, given the log-variance value as well as the information known at the current time.

The probability density of the log-variance $p_{\ln(v)}(u_t|u_s)$ is already given, and hence we need $p_{x|\ln(v)}(x_t|u_t, x_s, u_s)$. Although there is no closed-form expression for $p_{x|\ln(v)}$, we can derive its conditional characteristic function, $\phi(\omega; x_s, u_t, u_s)$, as

$$\begin{aligned}\phi(\omega; x_s, u_t, u_s) &= \mathbb{E}_s[\exp(i\omega x_t|u_t)] \\ &= \exp\left(i\omega\left[x_s + \mu(t-s) + \frac{\rho}{\sigma_v}(e^{u_t} - e^{u_s} - \kappa_v\theta_v(t-s))\right]\right) \\ &\quad \times\Phi\left(\omega\left(\frac{\kappa_v\rho}{\sigma_v} - \frac{1}{2}\right) + \frac{1}{2}i\omega^2(1-\rho^2); e^{u_t}, e^{u_s}\right),\end{aligned} \tag{6.8}$$

where $\Phi(u; v_t, v_s)$ is the characteristic function of the time integrated variance given by

$$\begin{aligned}\Phi(u; \nu_t, \nu_s) &= \mathbb{E}\left[\exp\left(iu\int_0^t \nu_\tau d\tau\right)\bigg|\nu_t, \nu_s\right] \\ &= \frac{I_q\left[\sqrt{\nu_t\nu_s}\frac{4\gamma(u)e^{-\frac{1}{2}\gamma(u)(t-s)}}{\eta^2(1-e^{-\gamma(u)(t-s)})}\right]}{I_q\left[\sqrt{\nu_t\nu_s}\frac{4\kappa_v e^{-\frac{1}{2}\kappa_v(t-s)}}{\eta^2(1-e^{-\kappa_v(t-s)})}\right]} \\ &\quad\times\frac{\gamma(u)e^{-\frac{1}{2}(\gamma(u)-\kappa_v)(t-s)}(1-e^{-\kappa_v(t-s)})}{\kappa_v(1-e^{-\gamma(u)(t-s)})} \\ &\quad\times\exp\left(\frac{\nu_s+\nu_t}{\eta^2}\left[\frac{\kappa_v(1+e^{-\kappa_v(t-s)})}{1-e^{-\kappa_v(t-s)}} - \frac{\kappa_v(1+e^{-\gamma(u)(t-s)})}{1-e^{-\gamma(u)(t-s)}}\right]\right),\end{aligned} \tag{6.9}$$

where $I_q(x)$ is the modified Bessel function of the first kind with order q. The variable $\gamma(u)$ is defined by

$$\gamma(u) = \sqrt{\kappa_v^2 - 2i\sigma_v^2 u}.$$

Under the dynamics of Heston with Merton jump-diffusion (2.1)–(2.2), the characteristic function of $\ln(S)$ is the multiplication of the characteristic function of $\ln(S)$ under Heston dynamics (6.1)–(6.2) and the characteristic function of the Merton jump-diffusion dynamics. Hence all the arguments below will apply to the SVJD dynamics (2.1)–(2.2).

The characteristic function of the jump-diffusion dynamics with log normal jump size can be calculated as

$$\phi_t^J(\omega) = \exp\{t\lambda(e^{-\delta^2\omega^2/2+i(\ln(k)-\frac{1}{2}\delta^2)\omega} - 1)\}. \tag{6.10}$$

6.2 The Pricing Method for European Options

In this section, we derive the COS formula for European-style options by replacing the density function by its Fourier cosine series. We make use of the fact that a density function tends to be smooth and therefore only a few terms in the expansion may already give a good approximation. Since the density $f(y|x)$ rapidly decays to zero as $y \to \infty$, we are able to truncate the infinite integration range without losing significant accuracy to $[a, b] \in \Re$, and we obtain approximation C_1, namely

$$C_1(x, t_0) = e^{-r\Delta t} \int_a^b g(y, T) f(y|x) dy. \tag{6.11}$$

Here, the idea is to approximate the probability density function of $x = \ln(S)$ by the Fourier cosine series, and we only need to recover the coefficients of the series. We will give more details on the choice of a and b in later sections.

In the second step, since $f(y|x)$ is usually not known whereas the characteristic function is, we replace the density by its cosine expansion in y, thus we have

$$f(y|x) = {\sum_{k=0}^{+\infty}}' A_k(x) \cos\left(k\pi\frac{y-a}{b-a}\right), \tag{6.12}$$

where $\sum'$ indicates that the first element in the summation is multiplied by one half, with

$$A_k(x) = \frac{2}{b-a} \int_a^b f(y|x) \cos\left(k\pi\frac{y-a}{b-a}\right) dy \tag{6.13}$$

so that

$$C_1(x,t_0) = e^{-r\Delta t}\int_a^b g(y,T)\sum_{k=0}^{+\infty}{}' A_k(x)\cos\left(k\pi\frac{y-a}{b-a}\right)dy. \tag{6.14}$$

We then interchange the summation and the integration and inserting the definition of V_k, we have

$$V_k = \frac{2}{b-a}\int_a^b g(y,T)\cos\left(k\pi\frac{y-a}{b-a}\right)dy \tag{6.15}$$

then we have

$$C_1(x,t_0) = \frac{b-a}{2}e^{-r\Delta t}\sum_{k=0}^{+\infty}{}' A_k(x)V_k. \tag{6.16}$$

Note that the V_k are the cosine series coefficients of payoff function $v(y,T)$ in y. Due to the rapid decay rate of these coefficients, we further truncate the series summation to obtain approximation C_2, thus we have

$$C_2(x,t_0) = \frac{b-a}{2}e^{-r\Delta t}\sum_{k=0}^{N-1}{}' A_k(x)V_k. \tag{6.17}$$

With the help of the characteristic function, the coefficients $A_k(x)$ defined in (6.13) can be approximated by

$$C(x,t_0) \approx C_3(x,t_0) = e^{-rt}\sum_{k=0}^{N-1}{}' Re\left(\phi\left(\frac{k\pi}{b-a};x\right)e^{-ik\pi\frac{a}{b-a}}\right)V_k. \tag{6.18}$$

where ϕ is the characteristic function. This is the COS formula for general underlying processes. We will show later that the V_k can be obtained analytically for plain vanilla and digital options, and that (6.18) can be simplified for the Heston model, so that many strikes can be handled simultaneously.

The key step in obtaining this semi-analytic formula (6.18) for option pricing is the replacement of the probability density function by its Fourier cosine series expansion. The advantage is that the product of the density and the payoff is transformed into a linear combination of products of cosine basis functions and a (payoff) function which is known analytically.

Important for convergence is therefore the convergence of the density function's cosine series, not the cosine series of the payoff, which appears only because we interchanged the summation and the integration in (6.16). Heuristically speaking, we decompose the probability density into a weighted sum of many "density-like basis functions" that can be used to calculate option values analytically. What matters for the accuracy and the computational speed is how well this probability density function is approximated.

Now the only part left to calculate is the coefficient V_k, which are the cosine coefficients of the payoff function $g(y)$, that is

$$V_k(T) = \begin{cases} V_k(0,b) \text{ call option}, \\ V_k(a,0) \text{ put option}, \end{cases} \tag{6.19}$$

where

$$V_k(l,u) = \frac{2}{b-a}\int_l^u g(y)\cos\left(k\pi\frac{y-a}{b-a}\right)dy. \tag{6.20}$$

Given that $g(x) = [\alpha \cdot K(e^x - 1)]^+$, we have

$$V_k(l,u) = \frac{2}{b-a}\alpha K[\chi_k(l^*,u^*) - \psi_k(l^*,u^*)],\ \alpha = \begin{cases} 1, \\ -1, \end{cases} \tag{6.21}$$

where $\alpha = 1$ for American call options and $\alpha = -1$ for American put options, and

$$l^* = \begin{cases} \max(l,0), & \alpha = 1, \\ \min(l,0), & \alpha = -1, \end{cases} \qquad u^* = \begin{cases} \max(u,0), & \alpha = 1, \\ \min(u,0), & \alpha = -1, \end{cases} \tag{6.22}$$

and

$$\chi_k(l^*,u^*) = \int_{l^*}^{u^*} e^x \cos\left(k\pi\frac{x-a}{b-a}\right)dx, \tag{6.23}$$

$$\psi_k(l^*,u^*) = \int_{l^*}^{u^*} \cos\left(k\pi\frac{x-a}{b-a}\right)dx. \tag{6.24}$$

Hence, χ_n, ψ_n are given by

$$\begin{aligned}\chi_k(l,u) = {} & \frac{1}{1+\left(\frac{k\pi}{b-a}\right)^2}\left[\cos\left(k\pi\frac{u-a}{b-a}\right)e^u - \cos\left(k\pi\frac{l-a}{b-a}\right)e^l\right. \\ & \left. + \frac{k\pi}{b-a}\sin\left(k\pi\frac{u-a}{b-a}\right)e^u - \frac{k\pi}{b-a}\sin\left(k\pi\frac{l-a}{b-a}\right)e^l\right]\end{aligned} \tag{6.25}$$

and

$$\psi_k(l,u) = \begin{cases} \left[\sin\left(k\pi\frac{u-a}{b-a}\right) - \sin\left(k\pi\frac{l-a}{b-a}\right)\right]\frac{b-a}{k\pi}, & k \neq 0, \\ u-l. & k = 0. \end{cases} \tag{6.26}$$

6.3 The Pricing Method for American Options

In this section, we derive the pricing formula for American options under the Heston model. This gives rise to a two-dimensional integral with a kernel which is only partly available in closed form. To evaluate this two dimensional integral, we develop a discrete formula based on Fourier cosine series expansions for the integration of the part of the kernel that is not known in closed form and a quadrature rule for the integral of the known part of the kernel. An efficient algorithm is used to compute the discrete formula with the help of the FFT algorithm.

6.3.1 *The pricing equations*

For a European option, which is defined at time s and matures at time t, with $0 < s < t$, the risk-neutral valuation formula reads as

$$C(x_s, u_s, s) = e^{-r(t-s)}\mathbb{E}^{\tilde{\mathbb{Q}}_s}[C(x_t, u_t, t)]. \tag{6.27}$$

Here $C(x_s, u_s, s)$ denotes the option price at time s, r is the risk-free interest rate, and $\mathbb{E}_s^{\tilde{Q}}$ is the expectation operator under the risk-neutral measure, $\mathbb{Q}$, given the information at s. The Markov property enables us to price an American option between two consecutive early exercise dates by the risk-neutral valuation formula (6.27). This value is then called the continuation value. The arbitrage-free price of the Bermudan option[1] on any early exercise date is the maximum of the continuation value and exercising the option. We are going to use a Bermudan option to approximate the American option. For a Bermudan option with M early exercise dates and $T = \{t_m, t_m < t_{m+1} | m = 0, 1, \ldots, M\}$, with $t_M \equiv T$ and $\Delta t := t_{m+1} - t_m$, the Bermudan option pricing formula reads as

$$C(x_{t_m}, u_{t_m}, t_m) = \begin{cases} g(x_{t_m}, t_m) & \text{for } m = M, \\ \max[Cont(x_{t_m}, u_{t_m}, t_m), g(x_{t_m}, t_m)] & \\ & \text{for } m = 1, 2, \ldots, M-1, \\ Cont(x_{t_m}, u_{t_m}, t_m) & \text{for } \quad m = 0, \end{cases} \tag{6.28}$$

where $g(x_t, t)$ is the payoff function at time t and $Cont(x_t, u_t, t)$ is the continuation value at time t. For simplicity, we use x_m and u_m for x_{t_m} and

[1]Bermuda options are exercisable at the date of expiration, and on certain number of specified dates that occur between the purchase date and the date of expiration.

u_{t_m}, respectively. The continuation value is given by

$$Cont(x_m, u_m, t_m) = e^{-r\Delta t}\mathbb{E}_{t_m}^{\tilde{Q}}[C(x_{m+1}, u_{m+1}, t_{m+1})], \tag{6.29}$$

which can be written as

$$Cont(x_m, u_m, t_m) = e^{-r\Delta t}\int_{\mathbb{R}}\int_{\mathbb{R}} C(x_{m+1}, u_{m+1}, t_{m+1}) \times p_{x,\ln(v)}(x_{m+1}, u_{m+1}|x_m, u_m)du_{m+1}dx_{m+1}. \tag{6.30}$$

From (6.7) we obtain

$$Cont(x_m, u_m, t_m) = e^{-r\Delta t}\int_{\mathbb{R}}\left[\int_{\mathbb{R}} C(x_{m+1}, u_{m+1}, t_{m+1})p_{x|\ln(v)}(x_{m+1}|u_{m+1}, x_m, u_m)dx_{m+1}\right] \times p_{\ln v}(u_{m+1}|u_m)du_{m+1}. \tag{6.31}$$

The inner integral in (6.31) equals the pricing formula for European options defined between t_m and t_{m+1}, provided the variance value at the future time point is known (keep in mind we are working backwards in time).

6.3.2 *Density recovery by Fourier cosine expansions*

The COS method, based on Fourier cosine expansions, is a very efficient method for the recovery of the probability density functions from the corresponding characteristic functions. It can therefore be efficiently used for the risk-neutral valuation formula in cases where the density is not known in closed form. We will apply the COS method to approximate the unknown conditional probability density, $p_{x|\ln(v)}$, in (6.31). The key idea of the COS method is to approximate the underlying probability density function, which is typically a smooth, real-valued function, by its Fourier cosine series expansion, taking into account the fact that the Fourier series coefficients have a direct connection to the characteristic function.

First we define a truncated integration range, $[a, b] \subset \mathbb{R}$, such that

$$\int_a^b p_{x|\ln(v)}(x_{m+1}|u_{m+1}, x_m, u_m)dx_{m+1} \leq TOL_x \tag{6.32}$$

for some predefined error tolerance TOL_x. This interval is defined as

$$[a, b] = [\xi_1 - 12\sqrt{|\xi_2|}, \xi_1 + 12\sqrt{|\xi_2|}], \tag{6.33}$$

where ξ_n denotes the $n-$th cumulant of the log-stock process. With an integration interval $[a, b]$ satisfying (6.33), we can recover the probability density from its Fourier cosine series expansion:

$$p_{x|\ln(v)}(x_{m+1}|u_{m+1},x_m,u_m) = \sum_{n=0}^{\infty}{}' G_n(u_{m+1},x_m,u_m)\cos\left(n\pi\frac{x_{m+1}-a}{b-a}\right). \tag{6.34}$$

The coefficients are the Fourier cosine coefficients, defined by

$$G_n(u_{m+1},x_m,u_m) = \frac{2}{b-a}\int_a^b p_{x|\ln(v)}(x_{m+1}|u_{m+1},x_m,u_m)\cos\left(n\pi\frac{x_{m+1}-a}{b-a}\right)dx_{m+1}.$$

In this case, the Fourier cosine expansion has some advantages as the series coefficients have a direct relation to the characteristic function and hence are known, in fact

$$G_n(u_{m+1},x_m,u_m) = \frac{2}{b-a}\Re\left\{\phi\left(\frac{n\pi}{b-a};x_m,u_{m+1},u_m\right)e^{-in\pi\frac{a}{b-a}}\right\}, \tag{6.35}$$

with $\phi(\theta;x,u_{m+1},u_m)$ is given by (6.8). By replacing G_n in (6.34) by (6.35) and truncating the series by N terms, we have a semianalytic formula which is able to approximate the probability density as well. So we can write

$$\begin{aligned} p_{x|\ln(v)}(x_{m+1}|u_{m+1},x_m,u_m) &= \sum_{n=0}^{N-1}{}'\frac{2}{b-a} \\ &\quad\times\Re\left\{\phi\left(\frac{n\pi}{b-a};x_m,u_{m+1},u_m\right)e^{-in\pi\frac{a}{b-a}}\right\}\cos\left(n\pi\frac{x_{m+1}-a}{b-a}\right) \\ &= \sum_{n=0}^{N-1}{}'\frac{2}{b-a} \\ &\quad\times\Re\left\{\phi\left(\frac{n\pi}{b-a};0,u_{m+1},u_m\right)e^{in\pi\frac{x_m-a}{b-a}}\right\}\cos\left(n\pi\frac{x_{m+1}-a}{b-a}\right). \end{aligned} \tag{6.36}$$

We can apply the fact $\phi(\theta;x_m,u_{m+1},u_m) = e^{i\omega x_m}\phi(\omega;0,u_{m+1},u_m)$, that is x_m can be be separated from the $u-$ terms and appears as a simple exponential term.

6.3.3 *Discrete Fourier-based pricing formula*

The option price at time t_0 can be recovered by recursion backwards in time. This is the same approach as in Fang and Oosterlee (2008) but here the integration is more complicated because of the two-dimensional integral.

Quadrature rule in the log-variance dimension First, we obtain the truncation range for the log-variance $[a_v, b_v]$ and then we are able to compute

$$Cont_1(x_m, u_m, t_m) = e^{-r\Delta t} \cdot \int_{a_v}^{b_v} \left[\int_a^b C(x_{m+1}, u_{m+1}, t_{m+1}) \right. \\ \left. \times p_{x|\ln(v)}(x_{m+1}|u_{m+1}, x_m, u_m) dx_{m+1} \right] \\ \times\, p_{\ln v}(u_{m+1}|u_m) du_{m+1}. \tag{6.37}$$

We use $Cont_i$ to track different approximations of the continuation value.

There are two ways to discretize the outer integral with respect to u_{m+1}, so that, by interpolation based quadrature rules or by a spectral series reconstruction of the interpolant (as in the COS method). Since $p_{ln(v)}$ itself is known analytically, we apply a J-point quadrature integration rule (like the GaussLegendre quadrature rule, the composite trapezoidal rule, trapezoidal rule or some similar method) to the outer integral, which gives

$$Cont_2(x_m, u_m, t_m) \\ = e^{-r\Delta t} \sum_{j=0}^{J-1} \omega_j \cdot p_{ln(v)}(\zeta_j | u_m) \\ \times \left[\int_a^b C(x_{m+1}, \zeta_j, t_{m+1}) p_{x|\ln(v)}(x_{m+1}|\zeta_j, x_m, u_m) dx_{m+1} \right]. \tag{6.38}$$

Here ω_j are the weights of the quadrature nodes ζ_j, $j = 0, 2, \ldots, J-1$

COS reconstruction in log-stock dimension Next, we need to replace $p_{x|\ln(v)}$ by the COS approximation and interchange the summation over n with the integration over x_{m+1} to obtain

$$Cont_3(x_m, u_m, t_m) = e^{-r\Delta t} \sum_{j=0}^{J-1} \omega_j \sum_{n=0}^{N-1}{}' V_{n,j}(t_{m+1}) \\ \Re\left\{ \tilde{\phi}\left(\frac{n\pi}{b-a}; \zeta_j, u_m\right) e^{in\pi \frac{x_m - a}{b-a}} \right\}, \tag{6.39}$$

where

$$V_{n,j}(t_{m+1})$$
$$= \frac{2}{b-a} \int_a^b C(x_{m+1}, u_{m+1}, t_{m+1}) \cos\left(n\pi \frac{x_{m+1} - a}{b-a}\right) dx_{m+1}, \tag{6.40}$$

and

$$\tilde{\phi}(\omega; u_{m+1}, u_m) = p_{\ln(v)}(u_{m+1}|u_m) \cdot \phi(\omega; 0, u_{m+1}, u_m). \tag{6.41}$$

The coefficients $V_{n,j}(t_{m+1})$ can be interpreted as the Fourier cosine series coefficients of the option value at time t_{m+1}. The expression $Cont_3$ (x_m, u_m, t_m) in (6.39) thus becomes a scaled inner product of the Fourier cosine series coefficients of the option price and of the underlying density. Finally, we interchange the summations in (6.39), which yields the discrete formula for the continuation value, namely

$$Cont_3(x_m, u_m, t_m) = e^{-r\Delta t} \Re \left\{ {\sum_{n=0}^{N-1}}' \beta_n(u_m, t_m) e^{in\pi \frac{x_m - a}{b-a}} \right\}, \tag{6.42}$$

where

$$\beta_n(u_m, t_m) = \sum_{j=0}^{J-1} \omega_j V_{n,j}(t_{m+1}) \tilde{\phi}\left(\frac{n\pi}{b-a}, \zeta_j, u_m\right). \tag{6.43}$$

Equation (6.42) expresses the continuation value at time t_m as a series expansion. The series coefficients, which depend only on the value of the variance (and not on the log-stock value) at time t_{m+1}, are (scaled) inner products of the cosine series coefficients of the option price at time t_{m+1} and the variance-dependent characteristic function of Φ.

Due to the use of a quadrature rule in the log-variance dimension, we compute on a log-variance grid. The same log-variance grid is employed for all time points, which gives

$$Cont_3(x_m, \zeta_l, t_m) = e^{-r\Delta t} \Re \left\{ {\sum_{n=0}^{N-1}}' \beta_n(\zeta_l, t_m) e^{in\pi \frac{x_m - a}{b-a}} \right\}, \tag{6.44}$$

with

$$\beta_n(\zeta_l, t_m) = \sum_{j=0}^{J-1} \omega_j V_{n,j}(t_{m+1}) \tilde{\phi}\left(\frac{n\pi}{b-a}, \zeta_j, \zeta_l\right). \tag{6.45}$$

For x_m, there is no computational grid needed since the price is constructed from a linear combination of cosine basis functions, in which the series coefficients do not depend on x_m itself. As such, x_m can be separated from the other variables; it is present only in the cosine functions. This enables us to derive an analytic formula for the series coefficients, as shown in the next subsection. One of the advantages of this spectral dimension is that the expression (6.42) is known for any value of $x_m \in R$, not just for discrete values. So, one can determine the early exercise points rapidly by solving

$$Cont_3(x_m, \zeta_j, t_m) - g(x_m) = 0,\ j = 0, 1, \ldots, J-1,$$

with an efficient root-finding procedure, such as Newton's method.

When the early exercise points, $x^*(u_m, t_m)$, have been determined, the procedure (6.28) can be used to compute the Bermudan option price. More specifically, the following hold:

- $C(x_M, u_M, t_M) = g(x_M)$, at maturity when $t = t_M$;

  $$\hat{C}(x_m, u_m, t_m) = \begin{cases} g(x_m) & \text{for } x \in [a, x^*(u_m, t_m)] \\ Cont_3(x_m, u_m, t_m) & \text{for } x \in [x^*(u_m, t_m), b] \end{cases} \quad (6.46)$$

  for a put option, and

- $$\hat{C}(x_m, u_m, t_m) = \begin{cases} Cont_3(x_m, u_m, t_m) & \text{for } x \in [a, x^*(u_m, t_m)] \\ g(x_m) & \text{for } x \in [x^*(u_m, t_m), b] \end{cases} \quad (6.47)$$

  for a call option.

- $\hat{C}(x_0, u_0, t_0) = Cont_3(x_0, u_0, t_0)$.

Here we use $\hat{C}$ to denote that we deal with approximate option values, due to the various approximations involved.

With the procedure above and expression (6.42), we can compute recursively $\hat{C}(x_0, u_0, t_0)$ from $\hat{C}(x_M, u_M, t_M)$ backwards in time. However, a more efficient technique exists. Instead of reconstructing $\hat{C}$ for each time point, we can recover the cosine series coefficients using backward recursion, and only at time t_0 do we apply (6.42) to reconstruct $\hat{C}$.

Backward recursion We show that the cosine coefficients of $\hat{C}(x_1, u_1, t_1)$ can be recovered recursively, with the FFT, from those of $\hat{C}(x_M, u_M, t_M)$ in $O((M-1) \times J \times N \times l)$ operations, with $l = \max[log2(N), J]$.

We first discuss the final time point, t_M. Since the option price at the maturity date equals the payoff (which does not depend on time), one can derive an analytic expression for $V_{n,j}(t_M)$ using (6.40), so that we can write

$$V_{n,j}(t_M) = \begin{cases} H_n(0,b) & \text{call option,} \\ H_n(a,0) & \text{put option,} \end{cases} \tag{6.48}$$

where the H_n- functions are the cosine coefficients of the payoff function $g(y)$, so that

$$H_n(l,u) = \frac{2}{b-a}\int_l^u g(y)\cos\left(n\pi\frac{y-a}{b-a}\right)dy. \tag{6.49}$$

Given that $g(x) = [\alpha \cdot K(e^x - 1)]^+$, we have

$$H_n(l,u) = \frac{2}{b-a}\alpha K[\chi_n(l^*,u^*) - \psi_n(l^*,u^*)],$$

$$\alpha = \begin{cases} 1 & \text{call option,} \\ -1 & \text{put option,} \end{cases} \tag{6.50}$$

where $\alpha = 1$ for American call options and $\alpha = -1$ for American put options, and where

$$l^* = \begin{cases} \max(l,0), & \alpha = 1, \\ \min(l,0), & \alpha = -1, \end{cases} \quad u^* = \begin{cases} \max(u,0), & \alpha = 1, \\ \min(u,0), & \alpha = -1, \end{cases} \tag{6.51}$$

and

$$\chi_n(l^*,u^*) = \int_{l^*}^{u^*} e^x \cos\left(n\pi\frac{x-a}{b-a}\right)dx, \tag{6.52}$$

$$\psi_n(l^*,u^*) = \int_{l^*}^{u^*} \cos\left(n\pi\frac{x-a}{b-a}\right)dx. \tag{6.53}$$

The quantities χ_n and ψ_n admit the analytic solutions

$$\begin{aligned}\chi_n(l,u) = \frac{1}{1+\left(\frac{n\pi}{b-a}\right)^2}\Bigg[&\cos\left(n\pi\frac{u-a}{b-a}\right)e^u - \cos\left(n\pi\frac{l-a}{b-a}\right)e^l \\ &+\frac{n\pi}{b-a}\sin\left(n\pi\frac{u-a}{b-a}\right)e^u - \frac{n\pi}{b-a}\sin\left(n\pi\frac{l-a}{b-a}\right)e^l\Bigg],\end{aligned} \tag{6.54}$$

and

$$\psi_n(l,u) = \begin{cases} \left[\sin\left(n\pi\frac{u-a}{b-a}\right) - \sin\left(n\pi\frac{l-a}{b-a}\right)\right]\frac{b-a}{n\pi}, & n \neq 0, \\ u-l, & n = 0. \end{cases} \tag{6.55}$$

Note that the applicability of the COS method is not limited to plain vanilla options. Analytic solutions for the Fourier cosine coefficients of binary options have also been obtained in Fang and Oosterlee (2008), so that discontinuous payoffs can also be dealt with highly efficiently.

Subsequently, we continue with time point t_{M-1}. By inserting $V_{n,j}(t_M)$ into (6.45), we obtain $\beta_n(\zeta_p, t_{M-1})$ for $p = 0, 1, \ldots, J-1$. With (6.44) one finds an analytic formula, $Cont_3(x_{M-1}, \zeta_p, t_{M-1})$, for the continuation value at time t_{M-1}. By Newton's method, we then solve $Cont_3(y, \zeta_p, t_{M-1}) - g(y) = 0$ to determine the location of the early exercise point, $y \equiv x^*(\zeta_p, t_{M-1})$. With the early exercise point, $x^*(\zeta_p, t_{M-1})$, known and $\hat{C}(x_{M-1}, \zeta_p, t_{M-1})$ as in (6.46) or (6.47), we split the integral in (6.40) into two parts (for $p = 0, 1, \ldots, J-1$)

$$\hat{V}_{k,p}(t_{M-1}) = \begin{cases} \hat{C}(x^*(\zeta_p, t_{M-1}), b, t_{M-1}) + H_k(a, x^*(\zeta_p, t_{M-1})) & \text{for a put,} \\ \hat{C}(a, x^*(\zeta_p, t_{M-1}), t_{M-1}) + H_k(x^*(\zeta_p, t_{M-1}), b) & \text{for a call,} \end{cases} \tag{6.56}$$

where $\hat{V}, \hat{C}$ denote approximate values. The $\hat{C}_{k,p}$ represent the cosine coefficients of the continuation value

$$\hat{C}_{k,p}(l, u, t_{M-1}) = \frac{2}{b-a} \int_l^u Cont_3(y, \zeta_p, t_{M-1}) \cos\left(k\pi \frac{y-a}{b-a}\right) dy. \tag{6.57}$$

After replacing $Cont_3$ in (6.57) by the COS approximation, interchanging summation and integration, we obtain

$$\hat{C}_{k,p}(l, u, t_{M-1}) = e^{-r\Delta t} \Re \left\{ {\sum_{n=0}^{N-1}}' \mathcal{M}_{k,n}(l, u) \beta_n(\zeta_p, t_{M-1}) \right\}, \tag{6.58}$$

with

$$\mathcal{M}_{k,n}(l, u) = \int_l^u \exp\left(in\pi \frac{y-a}{b-a}\right) \cos\left(k\pi \frac{y-a}{b-a}\right) dy, \tag{6.59}$$

which can be obtained analytically.

The expression above can be cast in an easily readable format in matrix/vector notation as

$$\hat{\boldsymbol{C}}(l, u, t_{M-1}) = e^{-r\Delta t} \Re\{\mathcal{M}(l, u) \boldsymbol{B}'(t_{M-1})\}, \tag{6.60}$$

where $\boldsymbol{B}'$ indicates that the first row of matrix $\boldsymbol{B}$ is multiplied by one half. The matrix $\mathcal{M}(l, u)$ is an $N \times N$ matrix composed of elements from $\mathcal{M}_{k,n}(l, u)$, and the matrix $\boldsymbol{B}(t_{M-1})$ is an $N \times J$ matrix, with J column vectors, so that

$$\boldsymbol{B}(t_{M-1}) = [\beta_0(t_{M-1}), \beta_1(t_{M-1}), \ldots, \beta_{J-1}(t_{M-1})]. \tag{6.61}$$

The column vectors (denoted by subscripts), $\beta_p(t_{M-1})$, are connected to the coefficients $V(t_M)$, so that the matrix with elements $V_{n,j}(t_M)$ can be written as

$$\beta_p(t_{M-1}) = [V(t_M) \cdot \tilde{\phi}(\zeta_p)]\boldsymbol{w}, \tag{6.62}$$

where $\boldsymbol{w}$ is a column vector (length J) with the quadrature weights and the (time-invariant) matrix $\tilde{\phi}(\zeta_p)$ is an $N \times J$ matrix with elements $\tilde{\phi}(\frac{n\pi}{b-a}, \zeta_j, \zeta_p)$, as defined in (6.41). The operator "$\cdot$" in the above equation denotes an elementwise matrix-matrix product.

From Fang and Oosterlee (2009) we know that matrix $\mathcal{M}(l,u)$ can be written as the sum of a Hankel matrix, $\mathcal{M}_c(l,u)$, and a Toeplitz matrix, $\mathcal{M}_s(l,u)$. Because matrix-vector products with Hankel and Toeplitz matrices can be transformed into circular convolutions of two vectors, the FFT algorithm can be applied to achieve the $O(Nlog_2(N))$ complexity in log-stock space.

In fact,

$$\mathcal{M}_{k,n}(l,u) = -\frac{i}{\pi}(\mathcal{M}^c_{k,n}(l,u) + \mathcal{M}^s_{k,n}(l,u)), \tag{6.63}$$

where

$$\mathcal{M}^c_{k,n}(l,u) = \begin{cases} \frac{(u-l)\pi i}{b-a} & k = n = 0, \\ \frac{\exp\left(i(n+k)\frac{(u-a)\pi}{b-a}\right) - \exp\left(i(n+k)\frac{(l-a)\pi}{b-a}\right)}{n+k} & \text{Otherwise;} \end{cases} \tag{6.64}$$

and

$$\mathcal{M}^s_{k,n}(l,u) = \begin{cases} \frac{(u-l)\pi i}{b-a} & k = n, \\ \frac{\exp\left(i(n-k)\frac{(u-a)\pi}{b-a}\right) - \exp\left(i(n-k)\frac{(l-a)\pi}{b-a}\right)}{n-k} & n \neq k. \end{cases} \tag{6.65}$$

We also know that the matrices, $\mathcal{M}_c = \{\mathcal{M}^c_{k,n}(l,u)\}_{k,n=0}^{N-1}$ and $\mathcal{M}_s = \{\mathcal{M}^s_{k,n}(l,u)\}_{k,n=0}^{N-1}$ are Hankel matrix and Toeplitz matrix respectively.

$$\mathcal{M}_c = \begin{bmatrix} m_0 & m_1 & m_2 & \cdots & m_{N-1} \\ m_1 & m_2 & \cdots & \cdots & m_N \\ \vdots & & & & \vdots \\ m_{N-2} & m_{N-1} & \cdots & \cdots & m_{2N-3} \\ m_{N-1} & m_N & \cdots & \cdots & m_{2N-2} \end{bmatrix}_{N\times N} \tag{6.66}$$

$$\mathcal{M}_s = \begin{bmatrix} m_0 & m_1 & m_2 & \cdots & m_{N-1} \\ m_{-1} & m_0 & m_1 & \cdots & m_{N-2} \\ \vdots & & & & \vdots \\ m_{2-N} & \cdots & m_{-1} & m_0 & m_1 \\ m_{1-N} & m_{2-N} & \cdots & m_{-1} & m_0 \end{bmatrix}_{N\times N} \tag{6.67}$$

with

$$m_j = \begin{cases} \frac{(u-l)\pi i}{b-a} & j = 0, \\ \frac{\exp\left(ij\frac{(u-a)\pi}{b-a}\right) - \exp\left(ij\frac{(l-a)\pi}{b-a}\right)}{j} & j \neq 0. \end{cases} \tag{6.68}$$

We require efficient algorithms for matrix-vector products, with a Toeplitz matrix, $\mathcal{M}_s$, and a Hankel matrix, $\mathcal{M}_c$. Due to the special structure of these matrices, we can rewrite these products into circular convolutions, that can be efficiently dealt with by the FFT algorithm. For Toeplitz matrices the product $\mathcal{M}_s u$ is equal to the first N elements of $\boldsymbol{m}_s \circledast \boldsymbol{u}_s$ with the $2N$-vectors

$$\boldsymbol{m}_s = [m_0, m_{-1}, m_{-2}, \ldots, m_{1-N}, 0, m_{N-1}, m_{N-2}, \ldots, m_1]^T$$

and $\boldsymbol{u}_s = [u_0, u_1, \ldots, u_{N-1}, 0, \ldots, 0]^T$.

For the Hankel matrix the product $\mathcal{M}_c u$ is equal to the first N elements of $\boldsymbol{m}_c \circledast \boldsymbol{u}_c$, in reversed order, with the $2N$-vectors:

$$\boldsymbol{m}_c = [m_{2N-1}, m_{2N-2}, \ldots, m_1, m_0]^T$$

and $\boldsymbol{u}_c = [0, \ldots, 0, u_0, u_1, \ldots, u_{N-1}]^T$.

The FFT algorithm can be employed since the circular convolution of two vectors is equal to the inverse discrete Fourier transform ($\mathcal{F}^{-1}$) of the products of the forward DFTs, $\mathcal{F}$, so that

$$\boldsymbol{x} \circledast \boldsymbol{y} = \mathcal{F}^{-1}\left\{\mathcal{F}\boldsymbol{x} \cdot \mathcal{F}\boldsymbol{y}\right\}.$$

Repeating the same computational procedure backwards in time, we can derive the equations that connect $\hat{V}(t_{m-1})$ to $\hat{V}(t_m)$, for $m = M-1, M-2, \ldots, 2$ and

$$\begin{cases} \hat{\boldsymbol{V}}(t_m) = \begin{cases} \hat{C}(x^*(\zeta_p, t_{m-1}), b, t_{m-1}) + H(a, x^*(\zeta_p, t_{m-1})) & \text{for a put,} \\ \hat{C}(a, x^*(\zeta_p, t_{m-1}), t_{m-1}) + H(x^*(\zeta_p, t_{m-1}), b) & \text{for a call,} \end{cases} \\ \hat{\beta}_j(t_{m-1}) = [\hat{V}(t_m) \cdot \tilde{\phi}(\zeta_j)]\boldsymbol{w}, \\ \hat{\boldsymbol{B}}(t_{m-1}) = [\beta_0(t_{m-1}), \beta_1(t_{m-1}), \ldots, \beta_{J-1}(t_{m-1})], \\ \hat{\boldsymbol{C}}(l, u, t_{m-1}) = e^{-r\Delta t}\Re\{\mathcal{M}(l, u)\boldsymbol{B}'(t_{m-1})\}. \end{cases} \tag{6.69}$$

We continue the procedure until $\hat{V}(t_1)$ has been recovered, which is then inserted into (6.42) and (6.45) to get a grid of option prices, $\hat{C}(x_0, \zeta_j, t_0)$, for $j = 0, 1, \ldots, J-1$. Now, one can either use a spline interpolation to get the value of $\hat{C}(x_0, u_0, t_0)$ from $\hat{C}(x_0, \zeta_j, t_0)$ or, at the initial stage of the computation, shift the u grid, so that u_0 lies exactly on the grid.

A number of remarks can be made about the above procedure:

- Special attention should be given to the calculation of $\phi(\omega, u_{m+1}, u_m)$. First, it involves a modified Bessel function of the first kind, which increases dramatically in value when $q \to 1$ and/or $\omega \to \infty$. Hence, the scaled Bessel function should be used instead. A robust package has been developed in Amos (1985, 1986) with algorithms to compute $I_d^*(z) = \exp(|Re\{z\}|)I_d(z)$ with a complex-valued argument, z, and a real-valued order d. As MATLAB (which we use here) incorporates this package for the MATLAB Bessel function, we replace $I_q(\cdot)$ by $e^{|Re\{\cdot\}|}I_q^*(\cdot)$ during the computations.
- The computation of the modified Bessel function costs significantly more (a factor of approximately 1000) CPU time than a simple multiplication, because the main part of the Bessel function algorithm is based on iterations. If the computation of the Bessel function costs $\mathcal{A}$ times the number of operations needed for a multiplication, a matrix based on $\tilde{\phi}(\frac{k\pi}{b-a}, \zeta_q, \zeta_j)$ would require $O(N \times J^2 \times \mathcal{A})$ operations to compute all matrix elements.
- If one employs equidistant quadrature rules for the log-variance dimension, then for a given value of k, the input argument of the Bessel function is a function of the grid point combination, $\zeta_q + \zeta_j$, which gives rise to the Hankel matrix (if ζ_j represents an equidistant grid). The favorable structure of a Hankel matrix enables us to determine only one row and one column of the $J \times J$ matrix for each value of k. The total number of operations needed is therefore reduced to $O(N \times J \times \mathcal{A})$. However, since the error convergence is much slower with equidistant quadrature rules, J should be set much larger than for Gaussian quadrature rules.

6.4 Two (Higher) Dimensional COS Methods

The two dimensional COS method presented here can be seen as an alternative (deterministic) pricing technique, which can deal with multiasset option problems of medium-sized dimensionality, meaning two-dimensional (2D) to approximately five-dimensional (5D) integrals involving the expectation operator. The method that we demonstrate for pricing higher-dimensional options is based on the Fourier transform of the transitional density function and is especially suitable for asset price models in the class of jump-diffusion processes. Detailed analysis on the convergence of the two dimensional COS methods together with other applications can be found in Ruijter and Oosterlee (2012).

Let $(\Omega, \mathcal{F}, P)$ be a probability space, $T > 0$ be a finite terminal time, and $\mathbb{F} = (\mathcal{F}_s)_{0 \leq s \leq T}$ be a filtration satisfying the usual conditions. The process $\boldsymbol{X}_t = (X_t^1, X_t^2)$ denotes a 2D stochastic process on the filtered probability space $(\Omega, \mathcal{F}, P)$, representing the log-asset prices. We assume that the bivariate characteristic function of the stochastic process is known, which is the case, for example, for affine jump-diffusions [9]. The value of a European rainbow option, with payoff function $g(\cdot)$, is given by the risk-neutral option valuation formula (2.1)

$$C(t_0, \boldsymbol{x}) = e^{-r\Delta t} \mathbb{E}^{t_0, \boldsymbol{x}}[g(\boldsymbol{X}_T)] = e^{-r\Delta t} \iint_{\mathbb{R}^2} g(\boldsymbol{y}) f(\boldsymbol{y}|\boldsymbol{x}) d\boldsymbol{y}.$$

Here, $\boldsymbol{x} = (x_1, x_2)$ is the current state, $f(y_1, y_2|x_1, x_2)$ is the conditional transition density function, r is the risk-free rate, and time to expiration is denoted by $\Delta t = T - t_0$. In the derivation of the COS formula, we distinguish three different approximation steps.

Step 1. We assume that the integrand is integrable, which is common for the problems we deal with. Because of that, we can, for given $\boldsymbol{x}$, truncate the infinite integration ranges to some domain $[a_1, b_1] \times [a_2, b_2] \in \mathbb{R}^2$ without losing significant accuracy. This gives the multidimensional Fourier cosine expansion formulation

$$\begin{aligned} C_1(t_0, \boldsymbol{x}) &= e^{-r\Delta t} \int_{a_2}^{b_2} \int_{a_1}^{b_1} g(\boldsymbol{y}) f(\boldsymbol{y}|\boldsymbol{x}) dy_1 dy_2 \\ &= e^{-r\Delta t} \int_{a_2}^{b_2} \int_{a_1}^{b_1} g(\boldsymbol{y}) \sum_{k_1=0}^{\infty}{}' \sum_{k_2=0}^{\infty}{}' A_{k_1,k_2}(\boldsymbol{x}) \\ &\quad \times \cos\left(k_1 \pi \frac{y_1 - a_1}{b_1 - a_1}\right) \cos\left(k_2 \pi \frac{y_2 - a_2}{b_2 - a_2}\right). \end{aligned} \tag{6.70}$$

The notation C_i is used for the different approximations of C and keeps track of the numerical errors that set in from each step. For final approximations we also use the "hat" notation, like $\hat{v}, \hat{c}$, etc. In the second line in (6.70), the conditional density is replaced by its Fourier cosine expansion in y on $[a_1, b_1] \times [a_2, b_2]$, with series coefficients $A_{k_1,k_2}(x)$ defined by

$$\begin{aligned} A_{k_1,k_2}(\boldsymbol{x}) &= \frac{2}{b_1 - a_1} \frac{2}{b_2 - a_2} \int_{a_2}^{b_2} \int_{a_1}^{b_1} f(\boldsymbol{y}|\boldsymbol{x}) \\ &\quad \times \cos\left(k_1 \pi \frac{y_1 - a_1}{b_1 - a_1}\right) \cos\left(k_2 \pi \frac{y_2 - a_2}{b_2 - a_2}\right) dy_1 dy_2. \end{aligned} \tag{6.71}$$

We interchange summation and integration and define

$$V_{k_1,k_2}(\boldsymbol{T}) = \frac{2}{b_1 - a_1}\frac{2}{b_2 - a_2}\int_{a_2}^{b_2}\int_{a_1}^{b_1} g(\boldsymbol{y}) \times \cos\left(k_1\pi\frac{y_1 - a_1}{b_1 - a_1}\right)\cos\left(k_2\pi\frac{y_2 - a_2}{b_2 - a_2}\right)dy_1dy_2, \quad (6.72)$$

which are the Fourier cosine series coefficients of $v(T, y) = g(y)$ on $[a_1, b_1] \times [a_2, b_2]$.

Step 2. Truncation of the series summations gives

$$C_2(t_0, \boldsymbol{x}) = \frac{b_1 - a_1}{2}\frac{b_2 - a_2}{2}e^{-r(T-t_0)} \times \sum_{k_1=0}^{N_1-1}{}' \sum_{k_2=0}^{N_2-1}{}' A_{k_1,k_2}(\boldsymbol{x})V_{k_1,k_2}(\boldsymbol{T}). \quad (6.73)$$

Step 3. Next, the coefficients $A_{k_1,k_2}(x)$ are approximated by

$$F_{k_1,k_2}(\boldsymbol{x}) = \frac{2}{b_1 - a_1}\frac{2}{b_2 - a_2}\iint_{\mathbb{R}^2} f(\boldsymbol{y}|\boldsymbol{x}) \times \cos\left(k_1\pi\frac{y_1 - a_1}{b_1 - a_1}\right)\cos\left(k_2\pi\frac{y_2 - a_2}{b_2 - a_2}\right)dy_1dy_2. \quad (6.74)$$

The 2D-COS formula is based on the following trigonometric relation

$$2\cos(\alpha)\cos(\beta) = \cos(\alpha + \beta) + \cos(\alpha - \beta), \quad (6.75)$$

with this, we have

$$2F_{k_1,k_2}(\boldsymbol{x}) = F^{+}_{k_1,k_2}(\boldsymbol{x}) + F^{-}_{k_1,k_2}(\boldsymbol{x}), \quad (6.76)$$

where

$$F^{\pm}_{k_1,k_2}(\boldsymbol{x}) = \frac{2}{b_1 - a_1}\frac{2}{b_2 - a_2}\iint_{\mathbb{R}^2} f(\boldsymbol{y}|\boldsymbol{x}) \times \cos\left(k_1\pi\frac{y_1 - a_1}{b_1 - a_1} \pm k_2\pi\frac{y_2 - a_2}{b_2 - a_2}\right)dy_1dy_2. \quad (6.77)$$

Now, the coefficients $F^{\pm}_{k_1,k_2}(\boldsymbol{x})$ can be calculated by

$$F^{\pm}_{k_1,k_2}(\boldsymbol{x}) = \frac{2}{b_1 - a_1}\frac{2}{b_2 - a_2}\Re\left(\phi\left(\frac{k_1\pi}{b_1 - a_1}, \frac{k_2\pi}{b_2 - a_2}|\boldsymbol{x}\right) \times \exp\left(-ik_1\pi\frac{a_1}{b_1 - a_1} \mp ik_2\pi\frac{a_2}{b_2 - a_2}\right)\right), \quad (6.78)$$

where $\phi(\cdot,\cdot|\boldsymbol{x})$ is the bivariate conditional characteristic function of $\boldsymbol{X}_T$, given $\boldsymbol{X}_{t0} = \boldsymbol{x}$. Here

$$\phi(\boldsymbol{u}|\boldsymbol{x}) = \mathbb{E}\left[e^{i\boldsymbol{u}\cdot\boldsymbol{X}_T}|\mathcal{F}_{t_0}\right] = \iint_{\mathbb{R}^2} e^{i\boldsymbol{u}\cdot\boldsymbol{y}} f(\boldsymbol{y}|\boldsymbol{x})d\boldsymbol{y}. \tag{6.79}$$

Insert (6.78) to (6.73), we will have

$$\begin{aligned}
\hat{C}(t_0,\boldsymbol{x}) &= \frac{b_1-a_1}{2}\frac{b_2-a_2}{2} \\
&\quad \times e^{-r\Delta t}\sum_{k_1=0}^{N_1-1}{}'\sum_{k_2=0}^{N_2-1}{}'\frac{1}{2}\left[F^+_{k_1,k_2}(\boldsymbol{x}) + F^-_{k_1,k_2}(\boldsymbol{x})\right]V_{k_1,k_2}(\boldsymbol{T}) \\
&= e^{-r\Delta t}\sum_{k_1=0}^{N_1-1}{}'\sum_{k_2=0}^{N_2-1}{}'\frac{1}{2}\left[\Re\left(\phi\left(\frac{k_1\pi}{b_1-a_1},\frac{k_2\pi}{b_2-a_2}|\boldsymbol{x}\right)\right.\right. \\
&\quad \left.\times\exp\left(-ik_1\pi\frac{a_1}{b_1-a_1} - ik_2\pi\frac{a_2}{b_2-a_2}\right)\right) \\
&\quad +Re\left(\phi\left(\frac{k_1\pi}{b_1-a_1},-\frac{k_2\pi}{b_2-a_2}|\boldsymbol{x}\right)\right. \\
&\quad \left.\left.\times\exp\left(-ik_1\pi\frac{a_1}{b_1-a_1} + ik_2\pi\frac{a_2}{b_2-a_2}\right)\right)\right]V_{k_1,k_2}(\boldsymbol{T}).
\end{aligned} \tag{6.80}$$

With the multidimensional-COS formula, calculation of the options Greeks is straightforward, as explained for the 1D case in Fang and Oosterlee (2008).

Remark 6.1. Cosine terms facilitate the usage of the characteristic function. Fourier sine expansions may also be used; however, their coefficients decrease at a lower rate for the payoff functions discussed, and because of this the cosine series are preferred. Alternative basis functions, like certain wavelet basis functions, may represent another interesting research direction for option pricing, but this is not yet known and is part of future research.

If the characteristic function is not available directly or not known analytically, it may be approximated. Local volatility models, for example, typically do not yield analytic functions ϕ, but recent research by Pagliarani *et al.* (2011) proposes a second-order approximation formula, so that an approximate characteristic function may be derived.

6.4.1 *American max option*

We generalize the multi-dimensional-COS method to pricing American max option. We denote a 2D underlying log-asset price process, $X_t = (X_{t1}, x_{t2})$, that is in the class of jump-diffusion processes. We approximate an American option by a Bermudan option with N early exercise times $t_0 < t_1 < \cdots t_m < \cdots < t_N = T$, with $\delta t := t_{m+1} - t_m$. The problem can be solved backward in time, with

$$\begin{cases} C(t_N, \boldsymbol{x}) = g(\boldsymbol{x}), \\ Cont(t_{m-1}, \boldsymbol{x}) = e^{-r\Delta t}\mathbb{E}[C(t_m, \boldsymbol{X}_{t_m})|\boldsymbol{X}_{t_{m-1}} = \boldsymbol{x}], \\ C(t_{m-1}, \boldsymbol{x}) = \max[g(\boldsymbol{x}), Cont(t_{m-1}, \boldsymbol{x})], 2 \le m \le N, \\ C(t_0, \boldsymbol{x}_0) = Cont(t_0, \boldsymbol{x}_0). \end{cases} \tag{6.81}$$

Function $Cont(t_{m-1}, \boldsymbol{x})$ is called the continuation value and is approximated by the 2D-COS formula,

$$Cont(t_{m-1}, \boldsymbol{x}) = \frac{b_1 - a_1}{2}\frac{b_2 - a_2}{2}e^{-r\Delta t}\sum_{k_1=0}^{N_1-1}{}'\sum_{k_2=0}^{N_2-1}{}' \times \frac{1}{2}[F^+_{k_1,k_2}(\boldsymbol{x}) + F^-_{k_1,k_2}(\boldsymbol{x})]V_{k_1,k_2}(\boldsymbol{T}). \tag{6.82}$$

The Fourier coefficients of the value function in (6.82) are given by

$$V_{k_1,k_2}(t_m) = \frac{2}{b_1 - a_1}\frac{2}{b_2 - a_2}\int_{a_2}^{b_2}\int_{a_1}^{b_1} C(t_m, \boldsymbol{y}) \times \cos\left(k_1\pi\frac{y_1 - a_1}{b_1 - a_1}\right)\cos\left(k_2\pi\frac{y_2 - a_2}{b_2 - a_2}\right)dy_1dy_2. \tag{6.83}$$

6.4.2 *Recursion formula for coefficients* $V_{k_1,k_2}(t_m)$

In this section, a recursive algorithm that recovers the coefficients $V_{k_1,k_2}(t_m)$, backwards in time, is derived.

When $m = N$, for some payoff function $g(y)$, there is an analytic formula for $V_{k_1,k_2}(t_N)$ but we do need to calculate it numerically for some other payoff function. The calculation of the coefficients depends on the continuation region and the early exercise region of the option.

For the coefficients that are used to approximate the continuation values at times $t_0, \ldots, t_{N-2}$, the value function, $C(t_m, \boldsymbol{x}) = \max[g(\boldsymbol{x}), Cont(t_m, \boldsymbol{x})]$, appears in the terms $V_{k1,k2}(t_m)$ and we need to find an optimal strategy for all state values $\boldsymbol{x} \in [a_1, b_1] \times [a_2, b_2]$. We divide the domain $[a_1, b_1] \times [a_2, b_2]$ into rectangular sub-domains $\mathcal{C}_q$ and $\mathcal{E}_p$, so that

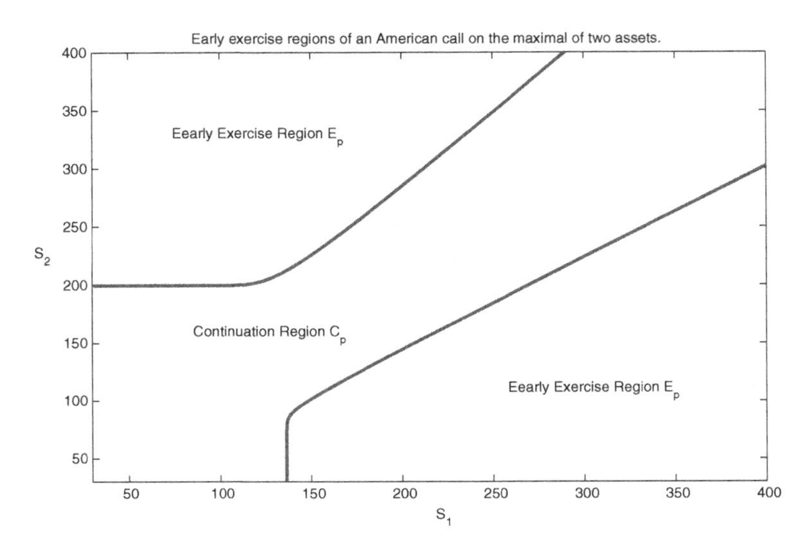

Fig. 6.1 Early exercise region of an American call on the maximum of two assets (see Broadie and Detemple (1997) for details).

for all states $\boldsymbol{x} \in \mathcal{C}_q$ it is optimal to continue and for all $\boldsymbol{x} \in \mathcal{E}_p$, it is optimal to exercise the option. The concept is demonstrated in Figure 6.1 for a call-on-maximum option. The early exercise regions $\mathcal{E}_p$ and the continuation regions $\mathcal{C}_q$ are labeled in the graph and the blue lines show an accurate boundary. We can split the integral in the definition of V_{k_1,k_2} into different parts. Thus we can write

$$\begin{aligned}
V_{k_1,k_2}(t_m) &= \frac{2}{b_1-a_1}\frac{2}{b_2-a_2}\sum_p \iint_{\mathcal{E}_p} g(\boldsymbol{y}) \\
&\quad \times \cos\left(k_1\pi\frac{y_1-a_1}{b_1-a_1}\right)\cos\left(k_2\pi\frac{y_2-a_2}{b_2-a_2}\right) d\boldsymbol{y} \\
&\quad + \frac{2}{b_1-a_1}\frac{2}{b_2-a_2}\sum_q \iint_{\mathcal{C}_q} Cont(t_m,\boldsymbol{y}) \\
&\quad \times \cos\left(k_1\pi\frac{y_1-a_1}{b_1-a_1}\right)\cos\left(k_2\pi\frac{y_2-a_2}{b_2-a_2}\right) d\boldsymbol{y} \\
&= \sum_p E_{k_1,k_2}(\mathcal{E}_p) + \sum_q C_{k_1,k_2}(t_m,\mathcal{C}_q)(m \neq 0, N). \qquad (6.84)
\end{aligned}$$

We approximate the terms $C_{k_1,k_2}(t_{N-1},[z_q,z_{q+1}]\times[\omega_q,\omega_{q+1}])$ in (6.84), where the variables z_q, z_{q+1}, ω_q and ω_{q+1} denote the corner points of the rectangular continuation region $\mathcal{C}_q$. For the integrand of the terms C_{k_1,k_2} we again apply the 2D Fourier cosine expansion by inserting the COS formula for $c(t_{N-1}, \boldsymbol{y})$, the approximation reads as

$$
\begin{aligned}
&\hat{C}_{k_1,k_2}(t_{N-1}, [z_q, z_{q+1}] \times [\omega_q, \omega_{q+1}]) \\
&= \frac{2}{b_1 - a_1} \frac{2}{b_2 - a_2} \int_{\omega_q}^{\omega_{q+1}} \int_{z_q}^{z_{q+1}} Cont(t_{N-1}, \boldsymbol{y}) \\
&\quad \times \cos\left(k_1 \pi \frac{y_1 - a_1}{b_1 - a_1}\right) \cos\left(k_2 \pi \frac{y_2 - a_2}{b_2 - a_2}\right) dy_1 dy_2 \\
&= \int_{\omega_q}^{\omega_{q+1}} \int_{z_q}^{z_{q+1}} e^{-r\Delta t} \sum_{j_1=0}^{N_1-1}{}' \sum_{j_2=0}^{N_2-1}{}' \\
&\quad \times \frac{1}{2} \left[F^+_{j_1,j_2}(\boldsymbol{x}) + F^-_{j_1,j_2}(\boldsymbol{x})\right] V_{j_1,j_2}(t_N) \\
&\quad \times \cos\left(k_1 \pi \frac{y_1 - a_1}{b_1 - a_1}\right) \cos\left(k_2 \pi \frac{y_2 - a_2}{b_2 - a_2}\right) dy_1 dy_2 \\
&= \Re\left(\sum_{j_1=0}^{N_1-1}{}' \sum_{j_2=0}^{N_2-1}{}' \frac{1}{2} e^{-r\Delta t} \phi\left(\frac{j_1\pi}{b_1 - a_1}, \frac{j_2\pi}{b_2 - a_2}\right)\right. \\
&\quad \left. \times V_{j_1,j_2}(t_N) M^+_{k_1,j_1}(z_q, z_{q+1}, a_1, b_1) M^+_{k_2,j_2}(\omega_q, \omega_{q+1}, a_2, b_2)\right) \\
&\quad + \Re\left(\sum_{j_1=0}^{N_1-1}{}' \sum_{j_2=0}^{N_2-1}{}' \frac{1}{2} e^{-r\Delta t} \phi\left(\frac{j_1\pi}{b_1 - a_1}, -\frac{j_2\pi}{b_2 - a_2}\right)\right. \\
&\quad \left. \times V_{j_1,j_2}(t_N) M^+_{k_1,j_1}(z_q, z_{q+1}, a_1, b_1) M^-_{k_2,j_2}(\omega_q, \omega_{q+1}, a_2, b_2)\right),
\end{aligned}
\tag{6.85}
$$

where the elements of square-matrices M^+ and M^- are given by

$$
M^+_{m,n}(u_1, u_2, a, b) = \frac{2}{b - a} \int_{u_1}^{u_2} e^{in\pi \frac{y-a}{b-a}} \cos\left(m\pi \frac{y - a}{b - a}\right) dy \tag{6.86}
$$

$$
M^-_{m,n}(u_1, u_2, a, b) = \frac{2}{b - a} \int_{u_1}^{u_2} e^{-in\pi \frac{y-a}{b-a}} \cos\left(m\pi \frac{y - a}{b - a}\right) dy. \tag{6.87}
$$

We thus find

$$\hat{C}_{k_1,k_2}(t_{N-1},[z_q,z_{q+1}]\times[\omega_q,\omega_{q+1}])$$
$$=\Re\left(\sum_{j_1=0}^{N_1-1}{}' M^{+}_{k_1,j_1}(z_q,z_{q+1},a_1,b_1)\mathcal{A}^q_{j_1,k_2}\right), \tag{6.88}$$

where

$$\begin{aligned}\mathcal{A}^q_{j_1,k_2} &= \sum_{j_2=0}^{N_2-1}{}'\frac{1}{2}e^{-r\Delta t}\phi\left(\frac{j_1\pi}{b_1-a_1},\frac{j_2\pi}{b_2-a_2}\right)\\ &\quad\times V_{j_1,j_2}(t_N)M^{+}_{k_2,j_2}(\omega_q,\omega_{q+1},a_2,b_2)\\ &\quad+\sum_{j_2=0}^{N_2-1}{}'\frac{1}{2}e^{-r\Delta t}\phi\left(\frac{j_1\pi}{b_1-a_1},-\frac{j_2\pi}{b_2-a_2}\right)\\ &\quad\times V_{j_1,j_2}(t_N)M^{-}_{k_2,j_2}(\omega_q,\omega_{q+1},a_2,b_2).\end{aligned} \tag{6.89}$$

The elements of $(N_1\times N_2)-$ matrix $\mathcal{A}^q$ are calculated in a row wise fashion. The row vector $\mathcal{A}^q_{j_1,\cdot}=\{\mathcal{A}^q_{j_1,k_2}\}_{k_2=0}^{N_2-1}$ can be written as two matrix-vector multiplications:

$$\mathcal{A}^q_{j_1,\cdot}=M^{+}_{k_2,j_2}(\omega_q,\omega_{q+1},a_2,b_2)\boldsymbol{w}^{q+}_{j_1,\cdot}+M^{-}_{k_2,j_2}(\omega_q,\omega_{q+1},a_2,b_2)\boldsymbol{w}^{q-}_{j_1,\cdot}, \tag{6.90}$$

where

$$\begin{aligned}\boldsymbol{w}^{q\pm}_{j_1,\cdot} &= \{w^{q\pm}_{j_1,j_2}\}_{j_2=0}^{N_2-1},\\ \text{with } w^{q\pm}_{j_1,j_2} &= \frac{1}{2}e^{-r\Delta t}\phi\left(\frac{j_1\pi}{b_1-a_1},\pm\frac{j_2\pi}{b_2-a_2}\right)V_{j_1,j_2}(t_N).\end{aligned} \tag{6.91}$$

Then the matrix $\hat{C}_{k_1,k_2}$ is computed in a column-wise fashion. The column-vector $\hat{C}_{\cdot,k_2}=\{\hat{C}_{k_1,k_2}\}_{k_1=0}^{N_1-1}$ is calculated by one matrix-vector product,

$$\hat{C}_{\cdot,k_2}(t_{N-1},[z_q,z_{q+1}]\times[\omega_q,\omega_{q+1}])=\Re(M^{+}(z_q,z_{q+1},a_1,b_1)\mathcal{A}^q_{\cdot,k_2}), \tag{6.92}$$

with column vector $\mathcal{A}^q_{\cdot,k_2}=\{\mathcal{A}^q_{j_1,k_2}\}_{j_1=0}^{N_1-1}$.

The coefficients $E_{k_1,k_2}([z_p,z_{p+1}]\times[\omega_p,\omega_{p+1}])$ are defined by

$$E_{k_1,k_2}([z_p,z_{p+1}]\times[\omega_p,\omega_{p+1}])$$

$$= \frac{2}{b_1 - a_1} \frac{2}{b_2 - a_2} \int_{\omega_p}^{\omega_{p+1}} \int_{z_p}^{z_{p+1}} g(\boldsymbol{y})$$

$$\times \cos\left(k_1 \pi \frac{y_1 - a_1}{b_1 - a_1}\right) \cos\left(k_2 \pi \frac{y_2 - a_2}{b_2 - a_2}\right) dy_1 dy_2. \tag{6.93}$$

These terms may admit an analytic solution, however in some practical applications an analytic solution may not exist. We need to develop methods to approximate these terms. Hence we end up with the approximate coefficients

$$\hat{V}_{k_1,k_2}(t_{N-1}) = \sum_p E_{k_1,k_2}(\mathcal{E}_p) + \sum_q C_{k_1,k_2}(\mathcal{C}_p). \tag{6.94}$$

For other coefficients $V_{k_1,k_2}(t_m)$, the approximation $Cont(t_m, \boldsymbol{y})$ and $V_{j_1,j_2}(t_{m+1})$, and the elements of the corresponding matrix $\mathcal{A}^q$ are

$$\mathcal{A}^q_{j_1,k_2} = \sum_{j_2=0}^{N_2-1}{}' \frac{1}{2} e^{-r\Delta t} \phi\left(\frac{j_1 \pi}{b_1 - a_1}, \frac{j_2 \pi}{b_2 - a_2}\right)$$

$$\times V_{j_1,j_2}(t_{m+1}) M^+_{k_2,j_2}(\omega_q, \omega_{q+1}, a_2, b_2)$$

$$+ \sum_{j_2=0}^{N_2-1}{}' \frac{1}{2} e^{-r\Delta t} \phi\left(\frac{j_1 \pi}{b_1 - a_1}, -\frac{j_2 \pi}{b_2 - a_2}\right)$$

$$\times V_{j_1,j_2}(t_{m+1}) M^-_{k_2,j_2}(\omega_q, \omega_{q+1}, a_2, b_2). \tag{6.95}$$

Remark 6.2. The matrix-vector products $M^+ v$ and $M^- v$ in the computation of matrices $\mathcal{A}^q$ and C can be computed efficiently by a Fourier-based algorithm. The computation time is $O(N \log_2 N)$, with N the length of the vector.

6.4.3 *Approximation methods for the coefficients* $\boldsymbol{V(T)}$ *and* $\boldsymbol{E(\mathcal{E}_p)}$

In this section, we develop methods for approximating the terminal coefficients $V_{k_1,k_2,\ldots,k_n}(T)$ and the terms $E_{k_1,k_2,\ldots,k_n}(\mathcal{E}_p)$ that are specific for the multidimensional-COS method.

In the 1D pricing problem, the terminal coefficients $V_{k_1}(T)$ admit analytic solutions for several options, like put- and call- based options, digital options, and power options. Besides, in the 1D-COS method for pricing Bermudan options, the terms $E_{k_1}(\mathcal{E}_p)$ are also usually known analytically.

In two dimensions, the payoff functions of, for instance, a geometric basket or max option provide analytic solutions to the 2D coefficients $V_{k_1,k_2}(T)$, but this is generally an exception. If no exact representation is available, then they can be approximated by using discrete cosine transforms (DCTs) or the Clenshaw-Curtis quadrature rule. The usage of DCTs is explained below. Also, analytic forms for the terms $E_{k_1,k_2,\ldots,k_n}(\mathcal{E}_p)$ are in general not available in the multidimensional version. An approximate method, based on Fourier cosine expansion of the payoff function, is discussed below as well.

DCTs. Next, we discuss the idea of using DCTs to approximate the terminal coefficients $V_{k_1,k_2}(T)$. For this we take $M \geq \max(N_1, N_2)$ grid-points for each spatial dimension and define

$$y_i^{n_i} = a_i + \left(n_i + \frac{1}{2}\right)\frac{b_i - a_i}{M}, \quad \text{and} \quad \Delta y_i = \frac{b_i - a_i}{M}, i = 1, 2. \tag{6.96}$$

The midpoint-rule integration gives us

$$\begin{aligned} V_{k_1,k_2}(T) &\approx \sum_{n_1=0}^{M-1}\sum_{n_2=0}^{M-1} \frac{2}{b_1 - a_1}\frac{2}{b_2 - a_2} g(y_1^{n_1}, y_2^{n_2}) \\ &\quad \times \cos\left(k_1\pi\frac{y_1^{n_1} - a_1}{b_1 - a_1}\right)\cos\left(k_2\pi\frac{y_2^{n_2} - a_2}{b_2 - a_2}\right)\Delta y_1 \Delta y_2 \\ &= \sum_{n_1=0}^{M-1}\sum_{n_2=0}^{M-1} g(y_1^{n_1}, y_2^{n_2}) \\ &\quad \times \cos\left(k_1\pi\frac{2n_1 + 1}{2M}\right)\cos\left(k_2\pi\frac{2n_2 + 1}{2M}\right)\frac{2}{M}\frac{2}{M}. \end{aligned} \tag{6.97}$$

The above 2D DCT refer to Type II can be calculated efficiently by, for example, the function dct2 of MATLAB. The approximated coefficients are denoted by $V_{k_1,k_2}^{DCT}(T)$, with the corresponding computed European option value $\hat{C}^{DCT}(t_0, \boldsymbol{x})$.

6.4.4 *3D-COS formula*

The 3D version of the COS formula reads as

$$\hat{C}(t_0, \boldsymbol{x}) = \prod_{i=1}^{3}\frac{b_i - a_i}{2} e^{-r\Delta t} \sum_{k_1=0}^{N_1-1}{}' \sum_{k_2=0}^{N_2-1}{}' \sum_{k_3=0}^{N_3-1}{}'$$

$$\times \frac{1}{4}[F^{++}_{k_1,k_2,k_3}(\boldsymbol{x}) + F^{+-}_{k_1,k_2,k_3}(\boldsymbol{x}) + F^{-+}_{k_1,k_2,k_3}(\boldsymbol{x})$$
$$+ F^{--}_{k_1,k_2,k_3}(\boldsymbol{x})] \times V_{k_1,k_2,k_3}(T). \tag{6.98}$$

The coefficients $F^{\pm\pm}_{k_1,k_2,k_3}(\boldsymbol{x})$ are given by the formula below with the characteristic function ϕ

$$\begin{aligned} F^{\pm\pm}_{k_1,k_2,k_3}(\boldsymbol{x}) &= \prod_{i=1}^{3} \frac{2}{b_i - a_i} \Re\left(\phi\left(\frac{k_1\pi}{b_1 - a_1}, \pm\frac{k_2\pi}{b_2 - a_2}, \pm\frac{k_3\pi}{b_3 - a_3}|\boldsymbol{x}\right)\right. \\ &\quad \left. \times \exp\left(-ik_1\pi\frac{a_1}{b_1 - a_1} \mp ik_2\pi\frac{a_2}{b_2 - a_2} \mp ik_3\pi\frac{a_3}{b_3 - a_3}\right)\right) \\ &= \prod_{i=1}^{3} \frac{2}{b_i - a_i} \Re\left(\phi_{\mathrm{JD}}\left(\frac{k_1\pi}{b_1 - a_1}, \pm\frac{k_2\pi}{b_2 - a_2}, \pm\frac{k_3\pi}{b_3 - a_3}\right)\right. \\ &\quad \left. \times \exp\left(ik_1\pi\frac{a_1}{b_1 - a_1} \pm ik_2\pi\frac{a_2}{b_2 - a_2} \pm ik_3\pi\frac{a_3}{b_3 - a_3}\right)\right), \end{aligned} \tag{6.99}$$

where the last equality holds for jump-diffusion processes, with $\phi_{\mathrm{JD}}(u_1, u_2, u_3) := \phi(u_1, u_2, u_3|0, 0, 0)$.

6.5 Numerical Results

To demonstrate the performance of the COS methods stated in this section and also be able to compare it to the method of lines algorithm outlined in Sec. 7.4 we implement the method for a given set of parameter values, chosen in order to best illustrate the impact that stochastic volatility and jump-diffusion may have on the early exercise boundary for an American call option. The parameter values used are listed in Table 6.1.

Table 6.1 Parameter values used for the American call option. The stochastic volatility (SV) parameters correspond to the Heston model. The jump-diffusion (JD) parameters correspond to the Merton model with log-normal jump sizes.

Parameter	Value	SV Parameter	Value	JD Parameter	Value
T	0.50	θ	0.04	λ^*	5.00
r	0.03	κ_v	2.00	γ	0.00
q	0.05	σ	0.40	δ	0.10
K	100	λ_v	0.00		
		ρ	± 0.50		

The prices of American options can be obtained by applying a Richardson extrapolation on prices of a few Bermudan options with small number of early exercise opportunities Chang *et al.* (2007), as demonstrated, for example, in Lord *et al.* (2008). Let $C(M)$ denote the value of a Bermudan option with M early exercise dates. We will use the following 4-point Richardson extrapolation scheme,

$$C_{AM}(d) = \frac{1}{21}\left(64C(2^{d+3}) - 56C(2^{d+2}) + 14C(2^{d+1}) - C(2^d)\right), \quad (6.100)$$

where $C_{AM}(d)$ denotes the approximated value of the American option. Now we price an American option using (6.100) with the 4-point Richardson extrapolation on Bermudan calls and vary the number of exercise dates.

Taking the results from PSOR as benchmark ("true" solution), we work out the RMSE of the COS method as shown in Table 6.2. Note that the (1K, 3K, 6K) refers to 1,000 steps in the time direction, 3,000 in the variance direction and 6,000 steps in the stock price direction. We refer the readers to Fang and Oosterlee (2011) for more analysis on the convergence of the method.

Table 6.2 American call prices computed using the COS method and Crank-Nicolson with PSOR (denoted PSOR). See Table 6.1 for the parameter values, with $\rho = -0.50$ and $v = 0.04$.

$\rho = -0.50$, $v = 0.04$ d in Eq. (6.100)	80	90	S 100	110	120	RMSRD (%)	Runtime (sec)
0	1.1365	3.3515	7.5932	13.8796	21.7159	0.0378	50
1	1.1363	3.3528	7.5943	13.8792	21.7165	0.0146	78
2	1.1361	3.3534	7.5972	13.8835	21.7185	0.0096	121
3	1.1357	3.3534	7.5973	13.8833	21.7187	0.0075	165
PSOR $(1 \times 10^3$, 3×10^3, $6 \times 10^3)$	1.1359	3.3532	7.5970	13.8830	21.7186	—	3,115,836

Chapter 7

A Numerical Approach to Pricing American Call Options under SVJD

In Chapter 3 we considered the simpler case of geometric Brownian motion plus jump-diffusion dynamics. In Chapter 4 we considered the case of stochastic volatility and jump-diffusion dynamics but in that case the volatility process was of the Heston type and the jumps were normally distributed. We showed, in particular how the transform method can be applied to this situation. In Chapter 5 we considered numerical approaches to the problem under Heston stochastic volatility dynamics. There we focused on the transform approach. In this chapter we consider the case, which combines stochastic volatility and jump-diffusion, as was postulated in Chapter 2 and more extensively in Chapter 4. This will allow a far more general structure to be investigated. In Sec. 7.1 we give some general considerations to the solution of the problem using various techniques and boundary solutions. In Sec. 7.2 we discuss the solution using the projected successive overrelaxation (PSOR) method to determine the "exact" solution. In Sec. 7.3 we discuss the componentwise splitting approach, which decomposes the discretised problem into three linear complementarity problems with tridiagonal matrices. These problems can be efficiently solved using the Brennan and Schwartz (1977, 1978) algorithm. The accuracy of the componentwise splitting (CS) method is increased by applying the Strang (1968) symmetrization. In Sec. 7.4, we give an outline of the method of lines solution to the problem. Finally, Sec. 7.5 discusses a numerical comparison between the different methods and shows that the method of lines (MOL) is indeed superior, probably because it calculates the price, the delta, the gamma and the free surface all at the same time.

7.1 The PDE Formulation

The American call $C^A(S,v,\tau)$ for a SVJD process of Chapter 2 can also be described with the complementarity form of a so-called obstacle problem

$$(\mathcal{L}C^A)(C^A - \phi) = 0$$
$$C^A - \phi \geq 0 \tag{7.1}$$

where $\mathcal{L}C^A$ is the Black-Scholes PIDE operator defined by (2.16), so that

$$\begin{aligned}\mathcal{L}C^A \equiv\ & \frac{vS^2}{2}\frac{\partial^2 C^A}{\partial S^2} + \rho\sigma_v vS\frac{\partial^2 C^A}{\partial S\partial v} + \frac{\sigma_v^2 v}{2}\frac{\partial^2 C^A}{\partial v^2} + (r - q - \lambda^* k^*)S\frac{\partial C^A}{\partial S} \\ & + (\alpha - \beta v)\frac{\partial C^A}{\partial v} - (r + \lambda^*)C^A \\ & + \lambda^* \int_0^\infty C^A(SY, v, \tau)G^*(Y)dY - \frac{\partial C^A}{\partial \tau}\end{aligned}$$

on $0 < S < \infty$, $0 < v < \infty$, $\tau \in (0,T]$. ϕ describes the obstacle on or above which C^A has to lie. The quantity ϕ is determined by the pay-off for the option. For a call it is

$$\phi(S,v,\tau) = S - K.$$

The initial condition imposed on (7.1) is the intrinsic value

$$C^A(S,v,0) = (S-K)^+.$$

If there are no jumps (that is, $\lambda^* = 0$) and if we also require that

$$\mathcal{L}C^A \leq 0, \tag{7.2}$$

which implies that $\mathcal{L}C^A \leq 0$ where $C^A = \phi$, then the well-established mathematical theory for obstacle problems (that is, the theory for variational inequalities Achdou and Pirroneau (2005)) guarantees that for smooth obstacles there exists a unique solution C^A of the obstacle problem. The solution is a continuously differentiable non-negative function $C^A(S,v,\tau)$ which satisfies (7.1) and (7.2). Moreover, C^A satisfies the PIDE

$$\mathcal{L}C^A = 0, \tag{7.3}$$

still with $\lambda = 0$, in the so-called continuation region, where C^A lies strictly above the obstacle, and

$$C^A(S,v,\tau) = S - K \tag{7.4}$$

in the so-called early exercise region where C^A coincides with the obstacle. On the boundary between these regions this C^A has to have continuous derivatives. This boundary does not specifically appear in the obstacle formulation. It is found a posteriori as the boundary of the set $\{S, v, \tau : C^A(S, v, \tau) > S - K\}$.

For the call it is commonly the case that the boundary is of the form

$$S = a(v, \tau)$$

and continuity of the derivatives translates into the value matching and smooth-pasting conditions

$$\begin{aligned} C^A(a(v,\tau), v, \tau) &= a(v,\tau) - K, \\ C^A_S(a(v,\tau), v, \tau) &= 1, \\ C^A_v(a(v,\tau), v, \tau) &= 0, \\ C^A_\tau(a(v,\tau), v, \tau) &= 0. \end{aligned} \tag{7.5}$$

Of relevance for numerical solutions is the observation that in general second derivatives of C^A are discontinuous on $a(v, \tau)$. The variational formulation of the obstacle problem suggests an application of finite elements as described, for example, in Achdou and Pirroneau (2005), which do not involve second derivatives.

As we saw in Chapter 2 the free boundary formulation of the American call consists of the PIDE (7.3) for $0 < S < a(v, \tau)$ and the free boundary conditions (7.5). Beyond $a(v, \tau)$ the call assumes its intrinsic value

$$C^A(S, v, \tau) = S - K.$$

The solution of the free boundary problem consists of $\{C^A(S, v, \tau), a(v, \tau)\}$, and even though the PIDE is a linear equation the resulting problem is inherently nonlinear.

When jumps are present, so that $\lambda^* > 0$, the equivalence of the complementarity and free boundary formulation of the call is rigorously established with PDE methods in Yang *et al.* (2006) for a constant volatility. We shall assume that the equivalence remains true for a stochastic volatility.

Here we shall consider only finite difference methods for the PIDE of the SVJD call, which we find more flexible and easier to implement than finite elements because of their simple data structure. They can be formulated for the complementarity form (7.1) and (7.2) or for the free boundary formulation (7.3) and (7.5). These approaches are quite dissimilar and will be the subject of the following sections. However, they all share the requirement that the integral in the pricing equation (7.3) be approximated. Although linear, it is a global term involving the unknown solution for all S. The term becomes local if we employ an iterative solution of (7.3). Let k be the iteration count. If C^{Ak-1} has been found then for $k \geq 1$ we replace the integral in (7.3) with the known source term

$$I^{k-1}(S,v,\tau) = \int_0^\infty C^{Ak-1}(SY,v,\tau)G(Y)dY$$

and solve the resulting complementarity or free boundary problem for C^{Ak}. The initial guess C^0 is not critical. In a time discrete approximation the solution from the preceding time step is a natural choice for the inital guess. All subsequent discussion of finite difference methods applies to the solution C^{Ak} of

$$\begin{aligned}\mathcal{L}_{BS}C^A \equiv & \frac{vS^2}{2}\frac{\partial^2 C^A}{\partial S^2} + \rho\sigma_v vS\frac{\partial^2 C^A}{\partial S\partial v} + \frac{\sigma_v^2 v}{2}\frac{\partial^2 C^A}{\partial v^2} + (r-q-\lambda^* k^*)S\frac{\partial C^A}{\partial S} \\ & + (\alpha - \beta v)\frac{\partial C^A}{\partial v} - (r+\lambda^*)C^A - \frac{\partial C^A}{\partial \tau} = I^{k-1}(S,v,\tau)\end{aligned} \tag{7.6}$$

where $0 < S < \infty$, $0 < v < \infty$, $\tau \in (0,T]$.

Any finite difference method for the American call with bounded grid size in S and v will necessarily have to be restricted to a finite computational domain

$$D = \{(S,v) : S_0 < S < S_{\max}, v_0 < v < v_{\max}\}$$

in order to obtain a finite-dimensional algebraic approximation. To make the problem solvable we need boundary data on the boundary ∂D of D. If $S_{\max}$ is sufficiently large so that it lies beyond the early exercise boundary $a(v,\tau)$ then we know that

$$C^A(S_{\max}, v, \tau) = S_{\max} - K. \tag{7.7}$$

Also, for sufficiently large $v_{\max}$ it is common to assume that the call value becomes insensitive to changes in the volatility. Hence we shall set

$$C_v^A(S, v_{\max}, \tau) = 0. \tag{7.8}$$

We remark that numerical simulations with the Heston model reported in Meyer (2015) suggest that a second order tangential boundary condition, a so called Venttsel boundary condition, may provide more consistent numerical prices and early exercise boundaries near $v_{\max}$ than (7.8), but also that the influence of the boundary condition at $v_{\max}$ dies out quickly as v decreases.

However, any boundary condition at S_0 and v_0 would have to be consistent with the behaviour of (7.6) at $S = 0$ and $v = 0$. Along these boundaries, the equation is a degenerate parabolic equation with given source term, and any time discrete approximation is a degenerate elliptic problem to which the discussion of Chapter 2 applies. As shown there the boundary condition (7.8) implies that

$$C^A(0, v, \tau) = 0. \tag{7.9}$$

If, in order to avoid the degeneracy of (7.6) at $S = 0$, we choose $0 < S_0 \ll 1$ we shall set

$$C^A(S_0, v, \tau) = 0.$$

(Alternatively, we can regularize (7.6) with the lead-off coefficient

$$\max\left\{\varepsilon, \frac{vS^2}{2}\right\}$$

and require (7.9). The simulations of Meyer (2015) typically use $\varepsilon = 10^{-4}$.)

At v_0 we shall impose the restriction of (7.6) to $v = 0$, i.e.

$$\begin{aligned}(r - q - \lambda^* k^*)SC_S^A(S, 0, \tau) + \alpha C_v^A(S, 0, \tau)\\ - (r + \lambda^*)C^A - C_\tau^A(S, 0, \tau) = I^{k-1}(S, 0, \tau),\end{aligned} \tag{7.10}$$

as boundary condition regardless of the algebraic sign of the Fichera function

$$h(S, 0) = \alpha - \frac{\sigma_v^2}{2}$$

because it is a legitimate oblique derivative boundary condition at every step of the iteration (see again Chapter 2).

However, in contrast to writing special numerical approximations to (7.10) at mesh points lying on the boundary $v = 0$ we shall simply approximate the call prices at $v = 0$ with a quadratic extrapolation of the latest computed call prices at the three mesh points v_1, v_2, v_3 nearest to $v_0 = 0$. Thus, for a constant mesh width Δv we shall always impose the boundary value

$$C^A(S, 0, \tau) = 3C^A(S, v_1, \tau) - 3C^A(S, v_2, \tau) + C^A(S, v_3, \tau). \tag{7.11}$$

The rationale for this approximation is the observation that if $c(v)$ is a smooth function defined on $[0, v_{\max}]$, and if the derivatives $c''(v)$ and $c'(v)$ at $v_1 = \Delta v$ are approximated by central difference quotients then a Taylor expansion shows that

$$c(0) - [3c(v_1) - 3c(v_2) + c(v_3)] = 0(\Delta v^3),$$

$$\frac{c(v_2)-c(0)}{2\Delta v} = \frac{c(v_2)-(3c(v_1)-3c(v_2)+c(v_3))}{2\Delta v}$$
$$\to c'(v_1) \to c'(0) \quad \text{as} \quad \Delta v \to 0$$

and

$$\Delta v\left[\frac{c(0)+c(v_2)-2c(v_1)}{\Delta v^2}\right]$$
$$= \Delta v\left[\frac{3c(v_1)-3c(v_2)+c(v_3)+c(v_2)-2c(v_1)}{\Delta v^2}\right]$$
$$= \Delta v\left[\frac{c(v_1)+c(v_3)-2c(v_2)}{\Delta v^2}\right] \to 0 \quad \text{as } \Delta v \to 0.$$

Hence if the unknown solution of (7.10) on $v = 0$ is approximated with the quadratic extrapolation of the latest computed interior prices then the difference equation at $v = v_1$ is a consistent approximation of the pricing equation (7.10) at $v = 0$ as $\Delta v \to 0$. The above extrapolation is simpler to implement than a finite difference approximation of (7.10) and appears to perform reliably.

For all numerical methods considered in this study the evaluation of the source term in (7.6) is independent of the algorithm employed to find $C^{Ak}(S_i, v_m, \tau_n)$. Known point values $\{C^{Ak-1}(S_i, v_m, \tau_n)\}_{i=0}^{I}$ are interpolated with a cubic spline $P_{m,n}^{k-1}(S)$ and the source term $I^{k-1}(S, v_m, \tau_n)$ in (7.6) is approximated by

$$I_{m,n}^{k-1}(S) = \int_0^\infty P_{m,n}^{k-1}(YS)G(Y)dY. \tag{7.12}$$

This integral is evaluated with a suitable quadrature rule. In particular, for lognormally distributed jump sizes we make the change of variable

$$X = [\ln Y - (p - \delta^2/2)/\sqrt{2}\delta$$

so that

$$I_{m,n}^{k-1}(S) = \frac{1}{\sqrt{\pi}}\int_{-\infty}^{\infty} e^{-X^2} P_{m,n}^{k-1}(S\exp\{(p-\delta^2/2)+\sqrt{2}\delta X\})dX.$$

The integral is approximated with a J point Hermite-Gauss quadrature scheme which yields the final approximation

$$I_{m,n}^{k-1}(S) \simeq \sum_{j=1}^{J} w_j^H P_{m,n}^{k-1}(S\exp\{(p-\delta^2/2)+\sqrt{2}\delta X_j^H\}) \tag{7.13}$$

where the abscissa X_j^H is the jth root of the Hermite polynomial $H_J(x)$ and the weights are given by

$$w_j^H = \frac{2^{J-1}J!}{J^2[H_{J-1}(x_j)]^2}.$$

We refer the reader to Abramowitz and Stegun (1970) for the computation of the weights and abscissas.

The main topic of this chapter is a comparison of four numerical methods for the approximate solution of the American call under SVJD which have been suggested in the literature. All four (including the MOL) were implemented at UTS and their performance compared.

Researchers with particular insight into any one method undoubtedly would be able to provide a more efficient code than ours for the method, but we think that our experience will remain applicable and help to see the methods in perspective. We begin our discussion with projected overrelaxation which is used to find reliable solutions against which those of other methods can be compared.

7.2 Numerical Solution using Finite Differences with Projected Overrelaxation (PSOR)

In this section we will no longer indicate the difference between C^A and C. We shall assume that C refers to C^A unless otherwise specified.

Projected overrelaxation is an iterative solution method for the algebraic complementarity problem which arises when the derivatives in (7.1) are replaced by difference quotients. The algorithm is almost as simple as a point relaxation method for a linear difference equation. Moreover, if the algebraic equation has the correct structure, its convergence is guaranteed. In this study PSOR is the inner iteration used to solve the algebraic equations approximating (7.6). However, since C^{k-1} is only an intermediate solution used to update the jump term, we carry out only one PSOR sweep through the mesh points and then update the integral term. Hence k is the iteration count for the successive PSOR sweeps. This combined inner/outer iteration is not known a-priori to converge, but is observed to do so. We go on the assumption that this PSOR approach on a sufficiently fine grid will yield highly accurate "benchmark" prices for the American call under SVJD.

We have implemented a finite difference PSOR to find the benchmark prices of the American options. The implementation of PSOR is detailed in this section.

The three variables S, v and τ are discretised as

$$S_i = i \times \Delta S, \quad i = 0, \ldots, I;$$
$$v_m = m \times \Delta v, \quad m = 0, \ldots, M;$$
$$\tau_n = T - n \times \Delta t, \quad n = 0, \ldots, N.$$

The options prices at those discrete points thus are

$$C^n_{i,m} = C(S_i, v_m, \tau_n),$$

with an initial condition:

$$C^0_{i,m} = (S_i - K)^+.$$

Similar to the discussion in Ekstrom *et al.* (2009), we use central differences to approximate most of the first and second derivatives in the PDE, except for forward and backward finite difference approximations on the boundaries. Thus we set,

$$\frac{\partial C}{\partial S} = \frac{C^n_{i+1,m} - C^n_{i-1,m}}{2\Delta S}, \quad \frac{\partial^2 C}{\partial S^2} = \frac{C^n_{i+1,m} - 2C^n_{i,m} + C^n_{i-1,m}}{\Delta S^2};$$
$$\frac{\partial C}{\partial v} = \frac{C^n_{i,m+1} - C^n_{i,m-1}}{2\Delta v}, \quad \frac{\partial^2 C}{\partial v^2} = \frac{C^n_{i,m+1} - 2C^n_{i,m} + C^n_{i,m-1}}{\Delta v^2};$$

and

$$\left.\frac{\partial C}{\partial S}\right|_{S=S_0} = \frac{C^n_{1,m} - C^n_{0,m}}{\Delta S}, \left.\frac{\partial C}{\partial S}\right|_{S=S_I} = \frac{C^n_{I,m} - C^n_{I-1,m}}{\Delta S},$$
$$\left.\frac{\partial C}{\partial v}\right|_{v=v_0} = \frac{C^n_{i,1} - C^n_{i,0}}{\Delta v}, \left.\frac{\partial C}{\partial v}\right|_{v=v_M} = \frac{C^n_{i,M} - C^n_{i,M-1}}{\Delta v}.$$

Our numerical results are obtained with the following approximations of the cross derivative. For $\rho \leq 0$ we use

$$\frac{\partial C^2}{\partial S \partial v} \approx \frac{1}{2}\left(\frac{C^n_{i+1,m+1} - C^n_{i,m+1} - (C^n_{i+1,m} - C^n_{i,m})}{\Delta S \Delta v} + \frac{C^n_{i,m} - C^n_{i-1,m} - (C^n_{i,m-1} - C^n_{i-1,m-1})}{\Delta S \Delta v}\right) \tag{7.14}$$

and for $\rho > 0$ we have

$$\frac{\partial C^2}{\partial S \partial v} \approx \frac{1}{2}\left(\frac{C^n_{i,m+1} - C^n_{i-1,m+1} - (C^n_{i,m} - C^n_{i-1,m})}{\Delta S \Delta v} + \frac{C^n_{i+1,m} - C^n_{i,m} - (C^n_{i+1,m-1} - C^n_{i,m-1})}{\Delta S \Delta v}\right). \tag{7.15}$$

At $v = 0$ the quadratic extrapolant is

$$C^n_{i,0} = 3C^n_{i,1} - 3C^n_{i,2} + C^n_{i,3}$$

and at $v = v_{\max}$ we enforce

$$\frac{C^n_{i,M} - C^n_{i,M-1}}{\Delta v} = 0.$$

Boundary conditions at the boundaries $S = 0$ and $S = S_{\max}$ for a call option are

$$C^n_{0,m} = 0, \quad C^n_{I,m} = S_{\max} - K, \quad m = 0, ..., M,$$

while those for a put option would be set as

$$P^n_{0,m} = K, \; P^n_{I,m} = 0.$$

The above spatial discretization leads to a semi-discrete equation for the continuation region which has the matrix representation

$$\frac{dC}{dt} + AC = b(t) \tag{7.16}$$

where C is the $(I-1)(M-1)$ vector for the unknown prices at the interior mesh points of the $S - v$ grid and b is a vector depending on the integral term and the boundary data.

The semi-discrete equation (7.16) is approximated over $[\tau_n, \tau_{n+1}]$ with the two-level scheme

$$\begin{aligned}(I + \theta\Delta\tau A)C^{n+1} + \theta b^{n+1} =& (I - (1-\theta)\Delta\tau A)C^n + (1-\theta)b^n, \\ & n = 0, \ldots, N-1,\end{aligned} \tag{7.17}$$

or, with the obvious identifications,

$$BC^{n+1} + e^{n+1} = EC^n + d^n.$$

The algebraic complementarity problem based on (7.16) at time level τ_{n+1} is

$$\begin{aligned}BC^{n+1} + e^{n+1} \geq EC^n + d^n, \quad C^{n+1} \geq \phi \\ (BC^{n+1} + e^{n+1} - EC^n - d^n)^\top (C^{n+1} - \phi) = 0.\end{aligned} \tag{7.18}$$

It is solvable with PSOR because the matrix B is strictly diagonally dominant for sufficiently small Δt, see Huang and Pang (1998). The initial guess for C^{n+1} is the solution C^n, with

$$C^0_{i,m} = (S_i - K)^+.$$

In PSOR iteration $k, C^{n+1,k}$, is the projected solution obtained from (7.18) where e^{n+1} depends on the point values of (7.12) obtained from the spline interpolants of $C^{n+1,k-1}$ along the lines of constant v.

Consistent with the common practice of avoiding possible oscillations that can occur when applying the Crank-Nicolson integration scheme to time-discretized partial differential equations with non-smooth initial/ boundary values, we have used a fully implicit method ($\theta = 1$) for the first three time steps before changing over to the second order Crank-Nicolson method where $\theta = 1/2$.

In order to accelerate the convergence of the PSOR, we select the over-relaxation parameter $\omega = 1.25$ in the algorithm. After the PSOR iteration has converged one additional SOR sweep without projection is executed. For every v_m the four nearest call values bracketing the intrinsic value are interpolated with a cubic polynolmial $P_m(S)$. The early exercise boundary $a(v_m, \tau_n)$ is the solution of $P_m(S) - (S - K) = 0$ which is found with Newton's method.

7.3 Componentwise Splitting for the SVJD Call

Componentwise splitting as applied by Ikonen and Toivanen (2007) to the Heston model adapts the concept of operator splitting introduced by Yanenko (1971). It provides an approximate solution of multi-dimensional parabolic equations with a sequence of one-dimensional problems. Originally known as the method of fractional steps, operator splitting can be applied to a wide range of complex initial/boundary value problems for evolution equations. For a broad review of the theory and application of operator splitting we refer to the survey paper of Glowinski *et al.* (2008).

The componentwise splitting of Ikonen and Toivanen (2007) will be applied here to the time continuous approximation of equation (7.6) for the SVJD call. We shall summarize the exposition of Ikonen and Toivanen (2007) and compare the performance of our implementation of the algorithm with alternative numerical methods.

From the mathematical perspective a desireable discretization of the cross derivative in (7.6) would lead to an algebraic approximation of (7.20) involving a so-called M matrix. As shown by Ikonen and Toivanen (2007) this is possible by relating the cross derivative to a directional derivative and a judicious choice of local finite difference mesh sizes. We shall find it convenient to paraphrase the arguments of Ikonen and Toivanen (2007) by separating the analytic and approximation aspects which allows an alternative discretization on a fixed mesh.

Let $\gamma = (\gamma_1(S, v), \gamma_2(S, v))$ be a unit vector field in the $S - v$ plane. Assuming smoothness of $C(S, v, t)$, the directional derivatives of C in the direction of γ are

$$\begin{aligned} D_\gamma C &= C_S\gamma_1 + C_v\gamma_2 \\ D_{\gamma\gamma} C &= (C_S\gamma_1 + C_v\gamma_2)_S\gamma_1 + (C_S\gamma_1 + C_v\gamma_2)_v\gamma_2 \\ &= C_{SS}\gamma_1^2 + 2C_{Sv}\gamma_1\gamma_2 + C_{vv}\gamma_2^2 \\ &\quad + C_S(\gamma_{1S}\gamma_1 + \gamma_{1v}\gamma_2) + C_v(\gamma_{2S}\gamma_1 + \gamma_{2v}\gamma_2). \end{aligned}$$

When we solve for C_{Sv} and substitute the resulting expression into (7.6) we obtain the equivalent pricing equation

$$\begin{aligned} \mathcal{L}C &\equiv a_{11}(S,v)C_{SS} + a_{12}(S,v)C_{\gamma\gamma} + a_{22}(S,v)C_{vv} \\ &\quad + b_1(S,v)C_S + b_2(S,v)C_v - (r+\lambda^*)C - C_t \\ &= I^{k-1}(S,v,t) \end{aligned}$$

where

$$a_{11} = \frac{vS^2}{2} - \rho\sigma vS\frac{\gamma_1}{2\gamma_2}, \quad a_{12} = \frac{\rho\sigma vS}{2\gamma_1\gamma_2}, \quad a_{22} = \frac{\sigma^2 v}{2} - \rho\sigma vS\frac{\gamma_2}{2\gamma_1}$$

$$b_1 = (r - q - \lambda^* k^*)S + d_1, \quad b_2 = (\alpha - \beta v) + d_2$$

with

$$d_j = -\rho\sigma vS\left[\frac{\gamma_{jS}\gamma_1 + \gamma_{jv}\gamma_2}{2\gamma_1\gamma_2}\right], j = 1, 2.$$

When $D_{\gamma\gamma}C$ is approximated by a central difference quotient then the algebraic approximation of the a_{12} term in (7.14) will lead to an M matrix A_{Sv} provided $\frac{\rho}{\gamma_1\gamma_2} \geq 0$ so that for $\rho \neq 0$ we require $\rho\gamma_1\gamma_2 > 0$.

The a_{11} and a_{12} terms combined with central difference approximations also will contribute M matrices A_S and A_v provided that $a_{11} \geq 0$ and $a_{22} \geq 0$.

The lower order terms b_1 and b_2 will provide M matrices if upwinding is used for C_S and C_v.

We observe that if we choose for γ the direction $(\rho S, \sigma)$ then

$$\begin{aligned} a_{11}(S,v) &= \frac{vS^2}{2}(1-\rho^2) \geq 0 \\ a_{22}(S,v) &= 0. \end{aligned}$$

Hence it is possible to choose γ such that the algebraic approximation of (7.6) at a given time step will lead to an M matrix. However, the above direction may not be optimal because it forces first order accurate upwinding everywhere onto the C_S and C_v terms.

If we do choose

$$\gamma(S,v) = \frac{(\rho S, \sigma)}{\sqrt{(\rho S)^2 + \sigma^2}}$$

then

$$\gamma_{1v} = \gamma_{2v} = 0 \quad \text{and} \quad \gamma_1 \gamma_{1S} + \gamma_2 \gamma_{2S} = 0.$$

Suppose now that we have a uniform fixed mesh with mesh points $\{S_i, v_m\}$ defined on $[S_0, S_{\max}] \times [0, v_{\max}]$. Let (S^*, v^*) be the closest intersection of the line through (S_i, v_m) with direction $(\rho S_i, \sigma)$ and the mesh lines $S = S_{i+1}$ and $v = v_{m+1}$. The directional derivative $D_{\gamma\gamma}C(S_i, v_m, t)$ can then be approximated with the central difference quotient

$$D_{\gamma\gamma}C(S_i, v_m, t_n) \\ = \frac{C(S_i + S^*, v_m + v^*, t_n) + C(S_i - S^*, v_m - v^*, t_n) - 2C(S_i, v_m, t_n)}{(\Delta S^{*2} + \Delta v^{*2})} \tag{7.19}$$

where $\Delta S^* = S^* - S_i, \quad \Delta v^* = v^* - v_m$.

In general, (S^*, v^*) will not be a mesh point of the uniform grid $\{S_i, v_m\}$. However, $C(S_i^*, v_m^*, t_n)$ can be approximated by a convex combination of the prices at the two nearest regular mesh points.

For example, if $0 < \Delta S^* \leq \Delta S = S_{i+1} - S_i$ so that $\Delta v^* = \Delta v$ we use the approximation

$$C(S_i + \Delta S^*, v_m + \Delta v^*, t_n) = \omega C(S_{i+1}, v_{m+1}, t_n) + (1 - \omega) C(S_i, v_{m+1}, t_n),$$

where $\omega = \Delta S^* / \Delta S$. Together with an analogous approximation for $C(S_i - \Delta S^*, v_m - \Delta v^*, t_n)$ we can thus replace (7.19) with a linear combination of central difference quotients for C_{vv} and the directional derivative along the vector $(\Delta S, \Delta v)$. When the approximation (7.19) is combined with central finite difference approximations for C_{SS} and C_{vv} and central or one-sided differences for $C_{\gamma\gamma}$ and C_γ we again obtain an ordinary differential equation

$$\frac{dC}{dt} + AC = b(t) \tag{7.20}$$

with an $IM \times IM$ matrix A which corresponds to a 7– point finite difference stencil at all interior mesh points.

If the interior mesh points $\{S_i, v_m\}$ are linearly ordered with index

$$q = (m-1)I + i, \quad 1 \leq i \leq I, \quad 1 \leq m \leq M$$

then the $M^*I \times M^*I$ matrix A in (7.20) is

$$A = A_S + A_{Sv} + A_v.$$

In the above ordering each of the component matrices is a tridiagonal matrix with

$$\begin{aligned}(A_S)_{qr} &= 0 \quad \text{for } r \neq q,\ q \pm 1,\\ (A_v)_{qr} &= 0 \quad \text{for } r \neq q,\ q \pm I,\\ (A_{Sv})_{qr} &= 0 \quad \text{for } r \neq q,\ q \pm (I+1) \quad \text{if } \rho > 0,\end{aligned}$$

and

$$(A_{Sv})_{qr} = 0 \quad \text{for } r \neq q,\ q \pm (I-1) \quad \text{if } \rho < 0.$$

As before, the right hand side $b(t)$ of (7.20) depends on the boundary data at S_0 and S_I, at v_0 and v_M and on the value of the integral (7.12) at the interior mesh points $\{S_i, v_m\}$.

If the solution $\{C^n\}$ of (7.20) has been found, then the analytic solution at time $\tau_{n+1} = \tau_n + \Delta\tau$ is

$$C^{n+1} = \exp(-A\Delta\tau)C^n + \int_{\tau_n}^{\tau_{n+1}} \exp(-A(\tau_{n+1} - s))b(s)ds.$$

In general the exponential of a matrix is costly to compute so that C^{n+1} is not readily obtainable. To efficiently obtain an approximate solution for C^{n+1}, a componentwise splitting is suggested in Ikonen and Toivanen (2007) for the evaluation of the matrix exponential. If A is split with the so-called Strang symmetrization

$$A = A_S + A_{Sv} + A_v = \left(\frac{A_S}{2} + \frac{A_v}{2} + A_{Sv} + \frac{A_v}{2} + \frac{A_S}{2}\right)$$

then it follows from the definition of the matrix exponential that in any vector norm

$$\| \exp[-A\Delta\tau] - \Phi(A) \| = 0(\Delta\tau^3)$$

where

$$\begin{aligned}\Phi(A) = {} & \exp[-A_S\Delta\tau/2]\exp[-A_v\Delta\tau/2]\exp[-A_{Sv}\Delta\tau]\\ & \times \exp[-A_v\Delta\tau/2]\exp[-A_S\Delta\tau/2].\end{aligned}$$

Moreover, if the integral is approximated with the trapezoidal rule

$$\int_{\tau_n}^{\tau_{n+1}} \exp(-A(\tau_{n+1} - s))b(S)ds = [b(\tau_{n+1}) + \exp(-A\Delta\tau)b(\tau_n)]\frac{\Delta\tau}{2}$$

then C^{n+1} is approximated to $O(\Delta\tau^3)$ with the formula

$$C^{n+1} = \Phi(A)\left[C^n + \frac{\Delta\tau}{2}b(\tau_n)\right] + \frac{\Delta\tau}{2}b(\tau_{n+1}). \tag{7.21}$$

Formula (7.21) will yield an approximate solution of a European option when formulated for a finite computational domain. It is quickly evaluated numerically. If we define

$$B_{S/2} = \left(I + \frac{\Delta\tau}{4}A_S\right), \quad C_{S/2} = \left(I - \frac{\Delta\tau}{4}A_S\right),$$

$$B_{v/2} = \left(I + \frac{\Delta\tau}{4}A_v\right), \quad C_{v/2} = \left(I - \frac{\Delta\tau}{4}A_v\right),$$

$$B_{Sv} = \left(I + \frac{\Delta\tau}{2}A_{Sv}\right), \quad C_{Sv} = \left(I - \frac{\Delta\tau}{2}A_{Sv}\right),$$

then as $\Delta\tau \to 0$

$$\| \exp[-A_S\Delta\tau/2] - B_{S/2}^{-1}C_{S/2} \| = O(\Delta\tau^3),$$

$$\| \exp[-A_v\Delta\tau/2] - B_{v/2}^{-1}C_{S/2} \| = O(\Delta\tau^3),$$

and

$$\| \exp[-A_{Sv}\Delta\tau] - B_{Sv}^{-1}C_{Sv} \| = O(\Delta\tau^3).$$

The first of these estimates implies that for any vector w^0,

$$\| \exp[-A_S\Delta\tau/2]w^0 - [B_{S/2}^{-1}C_{S/2}]w^0 \| = O(\Delta\tau^3).$$

If we set

$$w^1 = B_{S/2}^{-1}C_{S/2}w^0,$$

then w^1 is a solution of

$$B_{S/2}w^1 = C_{S/2}w^0.$$

Since $B_{S/2}$ is a tridiagonal matrix, w^1 is readily computed numerically. Hence if we set

$$w^0 = C^n + \frac{\Delta\tau}{2}b(\tau_n)$$

and compute

$$\begin{aligned}
B_{S/2}w^1 &= C_{S/2}w^0, \\
B_{v/2}w^2 &= C_{v/2}w^1, \\
B_{Sv}w^3 &= C_{Sv/2}w^2, \\
B_{v/2}w^4 &= C_{v/2}w^3, \\
B_{S/2}w^5 &= C_{S/2}w^4,
\end{aligned}$$

then to order $\Delta\tau^3$ the solution of (7.21) is given by

$$C^{n+1} = w^5 + \frac{\Delta\tau}{2}b(\tau_{n+1}).$$

Each w^j is the solution of a banded matrix system with two non-zero off-diagonals.

While the approximate solution $\{C^n\}$ is a second order solution of (7.20) over $[0, T]$, the intermediate solutions $\{w^1, w^2, w^3, w^4, w^5\}$ can also be interpreted as second order accurate solutions of the operator splitting sequence for the interval $[\tau_n, \tau_{n+1}]$

$$\frac{dc^1}{d\tau} + D_S c^1 = b(\tau), c^1(\tau_n) = c^n,$$

$$\frac{dc^2}{d\tau} + D_v c^2 = 0, c^2(\tau_n) = c^1(\tau_{n+1/2}), \tau_{n+1/2} = \tau_n + \Delta\tau/2,$$

$$\frac{dc^3}{d\tau} + D_{\gamma\gamma} c^3 = 0, c^3(\tau_n) = c^2(\tau_{n+1/2}),$$

$$\frac{dc^4}{d\tau} + D_v c^4 = 0, c^4(\tau_{n+1/2}) = c^3(\tau_{n+1}),$$

$$\frac{dc^5}{d\tau} + D_S c^5 = b(\tau), c^5(\tau_{n+1/2}) = c^4(\tau_{n+1})$$

so that

$$C^{n+1} = c^5(\tau_{n+1}).$$

In this view each c^i may be interpreted as a solution of the call problem in the direction of the differential operator. The numerical solution of c^i coincides with w^i for $i = 1, 2, 3, 4$. However, the approximation over the time step $\Delta\tau$ is only of order $\Delta\tau^2$. For example, the time continuous finite difference approximation of c^1 satisfies

$$\frac{dc^1}{d\tau} + A_S c^1 = b(\tau), \quad c^1(\tau_n) = C^n$$

so that

$$c^1(\tau_{n+1/2}) = \exp(-A\tau/2)C^n + \int_{\tau_n}^{\tau_{n+1/2}} \exp(-A(\tau_{n+1/2} - s))b(s)ds.$$

If we compare $c^1(\tau_{n+1/2})$ and w^1 then the integral has been approximated only to order $\Delta\tau^2$.

The same holds for $c^5(\tau_{n+1})$ because,

$$\left| c^5 - \left(w^5 + b(t_{n+1})\frac{\Delta\tau}{2} \right) \right| = O(\Delta\tau^2).$$

If this view is adopted then the application of componentwise splitting to the American call adds the requirement that all c^i and their approximations w^i must lie above the intrinsic value of the call. Since each w^i is found from a tridiagonal algebraic system, the Brennan Schwartz algorithm is recommended in Ikonen and Toivanen (2007) which insures that the solution does not fall below the intrinsic value of the call.

7.4 The Method of Lines for the SVJD Call

The application of the method of lines to pricing problems in finance is discussed in detail in a forthcoming monograph (see Meyer (2015)). Here we shall present only those aspects required for pricing puts and calls in the SVJD context.

In contrast to PSOR and componentwise splitting for the linear complementarity problem (7.1), the method of lines will be applied to the free boundary formulation of the call introduced in Chapter 2 and summarized at the beginning of this chapter.

The key idea behind the method of lines is to replace the PIDE with an equivalent system of one-dimensional integro-differential equations (IDEs), whose solution is readily obtained numerically. When volatility is constant, the system of IDEs is developed by discretising the time derivative. For the PIDE (2.16), we must also discretise the derivative terms involving the variance v, and adjust the algorithm for dealing with the integral term to cater for the fact that the option price and free surface also depend upon the variance v.

We begin by setting $v_m = m\Delta v$, where $m = 0, 1, 2, \ldots, M$. Typically we will set the maximum variance to be $v_M = 100\%$. Furthermore, we discretise the time to expiry according to $\tau_n = n\Delta\tau$, where $\tau_N = T$. We denote the option price along the variance line v_m and time line τ_n by $C(S, v_m, \tau_n) \equiv C_m^n(S)$, and set

$$V(S, v_m, \tau_n) = \frac{\partial C(S, v_m, \tau_n)}{\partial S} \equiv V_m^n(S), \tag{7.22}$$

which is of course the option delta at the grid point v_m.

We now select finite difference approximations for the derivative terms with respect to v. For the second order term, at the grid point (S, v_m, τ_n) we use the standard central difference scheme

$$\frac{\partial^2 C}{\partial v^2} = \frac{C_{m+1}^n - 2C_m^n + C_{m-1}^n}{(\Delta v)^2}, \tag{7.23}$$

and for the cross derivative term at the grid point (S, v_m, τ_n) we use the central difference approximation

$$\frac{\partial^2 C}{\partial S \partial v} = \frac{V_{m+1}^n - V_{m-1}^n}{2\Delta v}. \tag{7.24}$$

Since the coefficients of the second order derivative terms go to zero as $v \to 0$, we use an upwinding finite difference scheme (see Duffy (2006), Chapter 8) for the first order derivative term, such that, at the grid point (S, v_m, τ_n) we have

$$\frac{\partial C}{\partial v} = \begin{cases} \dfrac{C_{m+1}^n - C_m^n}{\Delta v} & \text{if } v \leq \dfrac{\alpha}{\beta}, \\ \dfrac{C_m^n - C_{m-1}^n}{\Delta v} & \text{if } v > \dfrac{\alpha}{\beta}. \end{cases} \tag{7.25}$$

Since the second order derivative terms both vanish as $v \to 0$, upwinding helps to stabilise the finite difference scheme with respect to v.

The integral term in (2.16) at each grid point is estimated during the iterative solution of the discretized equations as outlined in Sec. 7.1. It is obtained when the jump integral for the last available iterate of $C_m^n(S)$ (or its cubic spline interpolant) is evaluated with Hermite-Gauss quadrature (see (7.12). Thus we may assume that we have at all S the value $I_m^n(S)$ for the integral term.

Next we must select a discretisation for the time derivative. Initially we use a standard backward difference scheme, given at the grid point (S, v_m, τ_n) by

$$\frac{\partial C}{\partial \tau} = \frac{C_m^n - C_m^{n-1}}{\Delta \tau}. \tag{7.26}$$

This approximation is only first order accurate with respect to time. For the case of the standard American put option, Meyer and van der Hoek (1997) demonstrate that the accuracy of the method of lines increases considerably by using a second order approximation for the time derivative, specifically

$$\frac{\partial C}{\partial \tau} = \frac{3}{2}\frac{C_m^n - C_m^{n-1}}{\Delta \tau} - \frac{1}{2}\frac{C_m^{n-1} - C_m^{n-2}}{\Delta \tau}. \tag{7.27}$$

We initiate the method of lines solution by using (7.26) for the first three time steps, and then switch to (7.27) for all subsequent time steps.

Applying (7.13), (7.23)–(7.25) and the time discretisation to the PIDE (2.16), we must now solve a system of second order IDEs at each time step and variance grid point. For the first three time steps, the IDE at the grid point $v = v_m$ and $\tau = \tau_n$ is

$$\frac{v_m S^2}{2}\frac{d^2 C_m^n}{dS^2} + \rho\sigma v_m S \frac{V_{m+1}^n - V_{m-1}^n}{2\Delta v} + \frac{\sigma^2 v_m}{2}\frac{C_{m+1}^n - 2C_m^n + C_{m-1}^n}{(\Delta v)^2}$$

$$+\frac{\alpha-\beta v}{2}\frac{C_{m+1}^{n}-C_{m-1}^{n}}{\Delta v}+\frac{|\alpha-\beta v|}{2}\frac{C_{m+1}^{n}-2C_{m}^{n}+C_{m-1}^{n}}{\Delta v}$$
$$+(r-q-\lambda^{*}k^{*})S\frac{dC_{m}^{n}}{dS}-(r+\lambda^{*})C_{m}^{n}+\lambda^{*}I_{m}^{n}-\frac{C_{m}^{n}-C_{m}^{n-1}}{\Delta\tau}=0, \tag{7.28}$$

and for all subsequent time steps the IDE is

$$\frac{v_{m}S^{2}}{2}\frac{d^{2}C_{m}^{n}}{dS^{2}}+\rho\sigma v_{m}S\frac{V_{m+1}^{n}-V_{m-1}^{n}}{2\Delta z}+\frac{\sigma^{2}v_{m}}{2}\frac{C_{m+1}^{n}-2C_{m}^{n}+C_{m-1}^{n}}{(\Delta v)^{2}}$$
$$+\frac{\alpha-\beta v}{2}\frac{C_{m+1}^{n}-C_{m-1}^{n}}{\Delta v}+\frac{|\alpha-\beta v|}{2}\frac{C_{m+1}^{n}-2C_{m}^{n}+C_{m-1}^{n}}{\Delta v}$$
$$+(r-q-\lambda^{*}k^{*})S\frac{dC_{m}^{n}}{dS}-(r+\lambda^{*})C_{m}^{n}+\lambda^{*}I_{m}^{n}$$
$$-\frac{3}{2}\frac{C_{m}^{n}-C_{m}^{n-1}}{\Delta\tau}+\frac{1}{2}\frac{C_{m}^{n-1}-C_{m}^{n-2}}{\Delta\tau}=0. \tag{7.29}$$

We require two boundary conditions in the v direction, one at v_0 and the other at v_M. We set $\partial C/\partial v=0$ along the variance boundary $v=v_M$. When v is zero, we again fit a quadratic polynomial through the option prices at v_1, v_2 and v_3, and then use this to extrapolate an approximation of the price at v_0 to be used in (7.29) at $m=1$.

After taking the boundary conditions into consideration, at each time step n we must solve a system of $M-1$ second order IDEs along the variance lines. This is done using a two stage iterative scheme. We treat the IDEs as ODEs by using C_m^{n-1} as an initial approximation for C_m^n in the integral term I_m^n. We then solve the ODEs for increasing values of v with a line Gauss-Seidel iteration using the latest available estimates for C_{m+1}^n, C_{m-1}^n, V_{m+1}^n and V_{m-1}^n. The initial estimates for C_m^n and V_m^n are simply C_m^{n-1} and V_m^{n-1}. Otherwise we use the latest estimates for C_m^n and V_m^n found during the current iteration through the variance lines. We iterate until the price profile converges to a desired level of accuracy. Second, once the price has converged, we update the estimate of the integral term I_m^n using the current price profile estimate, and repeat the process until convergence is obtained for both levels of iteration. We then proceed to the next time step.

The generic first order form for (7.28) and (7.29) for fixed m is the system of two scalar equations

$$\frac{dC_m^n}{dS} = V_m^n, \tag{7.30}$$

$$\frac{dV_m^n}{dS} = A_m(S)C_m^n + B_m(S)V_m^n + G_m^n(S), \tag{7.31}$$

where $G_m^n(S)$ is a function of C_{m+1}^n, C_{m-1}^n, V_{m+1}^n, V_{m-1}^n, C_m^{n-1}, C_m^{n-2} and I_m^n. We solve (7.30)–(7.31) using the Riccati transformation.[1] Note that we are only able to apply the Riccati transformation to the system (7.30)–(7.31) provided that both equations are treated as ODEs. This is made possible by approximating I_m^n using the values C_m^{n-1} of the previous time step, as stated earlier, and then using an iterative technique in which the integral term is updated until the price converges. For later use we note that the right hand sides of equations (7.30) and (7.31) respectively yield the delta and gamma at each point of the grid.

The Riccati transformation is given by

$$C_m^n(S) = R_m(S)V_m^n(S) + W_m^n(S), \tag{7.32}$$

where R and W are solutions to the initial value problems

$$\frac{dR_m}{dS} = 1 - B_m(S)R_m(S) - A_m(S)(R_m(S))^2, \quad R_m(0) = 0, \tag{7.33}$$

$$\frac{dW_m^n}{dS} = -A_m(S)R_m(S)W_m^n - R_m(S)G_m^n(S), \quad W_m^n(0) = 0, \tag{7.34}$$

and V is the solution to

$$\frac{dV_m^n}{dS} = A_m(S)(R(S)V + W_m^n(S)) + B_m(S)V + G_m^n(S), \tag{7.35}$$

and

$$V_m^n(a_m^n) = 1,$$

where we denote the free boundary at grid point (v_m, τ_n) by $a(v_m, \tau_n) = a_m^n$.[2] Since R_m is independent of τ, we begin by solving (7.33) and storing

[1] The Riccati transformation basically replaces a given differential system (here equation (7.30) and equation (7.31)) with an equivalent set of uncoupled equations of lower dimension (here equation (7.33), equation (7.34) and equation (7.35) above).

[2] All ODEs have been solved by use of the implicit trapezoidal rule, discussed for example by Shampine (1994).

the solution. Next we solve (7.34) for increasing values of S, ranging from $0 < S < S_{\max}$, where we select $S_{\max}$ sufficiently large such that $S_{\max} > a_m^n$ will be guaranteed. We then step forward in S using the generated values of R_m and W_m^n until we encounter the value S^* such that[3] $C_m^n(S)$ given by (7.32) satisfies the value matching and smooth-pasting conditions at S^*, so that

$$S^* - K = R_m(S^*) + W_m^n(S^*), \tag{7.36}$$

S^* is the value of the free boundary at grid point (v_m, τ_n).[4] Once a_m^n has been determined we then solve (7.35) starting at $S = a_m^n$ and sweeping back to $S = 0$. Finally we use the calculated values of R_m, W_m^n and V_m^n in (7.32) to determine the option price at each grid point along the variance lines at time to maturity τ_n.

An alternative way of solving the second order scalar equation (7.28) or (7.29) for fixed m is with a finite difference method on a discrete grid $\{S_i\}$. The resulting linear algebraic equation involves a tridiagonal matrix which can quickly be solved with an implementation of Gaussian elimination known as the Thomas algorithm. It is known that the Thomas algorithm converges to the Riccati transformation as $\max_i(S_{i+1} - S_i) \to 0$. Loosely speaking, the Riccati transformation is the closure of Gaussian elimination as $\Delta S \to 0$ in the finite difference approximation of (7.28). Conversely, the numerical solution of the equations of the Riccati transformation corresponds to a finite difference solution for (7.28). Solving the free boundary problem for (7.28) or (7.29) with the trapezoidal rule applied to the Riccati transformation and searching for a root of (7.36) is roughly equivalent in complexity and efficiency to the Brennan Schwartz method for (7.28) for the call along each variance line (for more details see Meyer (2015)).

7.5 Numerical Results

Because the method of lines explicitly computes the early exercise surface $a(v, \tau)$ and the delta of the option, we have used it extensively for the simulation of the SVJD call.

[3]We test equation equation (7.36) at each grid point and find the grid points between which $S - K - R_m(S) - W_m^n(S)$ changes sign. We then use Newton's method to search for the value of S^* by fitting a cubic spline through four points bracketing S^*.

[4]We remind the reader that at S^* the first of the free boundary conditions (2.15) becomes $V_m^n(S^*) = 1$.

Table 7.1 Parameter values used for the American call option. The stochastic volatility (SV) parameters correspond to the Heston model. The jump-diffusion (JD) parameters correspond to the Merton model with log-normal jump sizes.

Parameter	Value	SV Parameter	Value	JD Parameter	Value
T	0.50	θ	0.04	λ^*	5.00
r	0.03	κ_v	2.00	γ	0.00
q	0.05	σ	0.40	δ	0.10
K	100	λ_v	0.00		
		ρ	± 0.50		

To demonstrate the performance of the method of lines algorithm outlined in Sec. 7.4 we implement the method for a given set of parameter values, chosen in order to best illustrate the impact that stochastic volatility and jump-diffusion may have on the early exercise boundary for an American call option. The parameter values used are listed in Table 7.1.

We consider the case where $r < q$, and a time to maturity of 6 months, as this best demonstrates the changes that arise in the free boundary when jumps are introduced. The value of σ is chosen intentionally large in order to emphasize the impact of stochastic volatility on the free boundary. We assume that jump sizes are log-normally distributed about a mean value of $Y = 1$ so that in a sense the jumps up and down average out. This allows us to focus on the impact that the Wiener correlation, ρ, has on the free boundary. In addition, the small value of δ has been chosen so as to avoid further increases to the overall variance of S.

When implementing the method of lines we take the following case as an example to show its convergence pattern. We use $N = 50$ time-steps and $M = 100$ volatility lines, with maximum volatility $v_M = 100\%$. We take a non-uniform grid in S, splitting the domain into three intervals. Given that the strike price is $K = 100$, the maximum value for S is set to 400, with a total of 1138 grid points (denoted by S_{pts}), distributed between the three intervals such that there are 40 points for $0.5 \le S \le 1$, there are 198 points for $1 \le S \le 100$, and finally 900 grid points for $100 \le S \le 400$.

For the Hermite-Gauss quadrature scheme in (7.13) we use $J = 50$ abscissa points. All iterative calculations utilise the stopping condition that the maximum over all S of $|C_m^n(S)^{(k)} - C_m^n(S)^{(k-1)}|$ is less than 10^{-8}, where the subscript k denotes the solution at the kth iteration. We note that for the parameter values given in Table 7.1, the solution along the volatility lines typically converges for less than 85 iterations, and convergence with

Table 7.2 Sample convergence pattern for the method of lines iterative procedures. Parameter values are as given in Table 7.1.

Integral Iteration	Volatility Iterations
1st	82
2nd	61
3rd	39
4th	18
5th	3
6th	1
Total SV iterations	204

Table 7.3 The changes of the number of required "volatility iterations" with respect to the changes of the number of points in the v direction.

No. of points in the v direction	No. of total volatility iterations
30	41
50	75
60	96
90	174
100	204
120	274

respect to updating the integral term (7.12) generally needs no more than 6 iterations. Furthermore, the number of iterations along the volatility lines reduce by more than a quarter after the first time the integral term is updated and continues to reduce by more than a half each time the integral term is updated. A typical sequence is provided in Table 7.2. It is also of interest to observe how the total number of volatility iterations changes as the number of volatility lines changes, this effect is shown in Table 7.3. It seems that the number of volatility iterations increases faster than the number of lines in the v direction. This deterioration in convergence is similar to that observed in a standard Gauss-Seidel scheme, and is probably due to the fact that as in Gauss-Seidel iteration we are using the latest updates for the values as we step through the (τ, v) grid. It is also worth noting that of all the components within the iterative scheme, computing the integral term (7.13) is the most computationally intensive, since we must perform J extrapolations of $C(S, v, \tau)$ with respect to S at every point in the S-v grid. Thus we are required to fit a total of M cubic splines

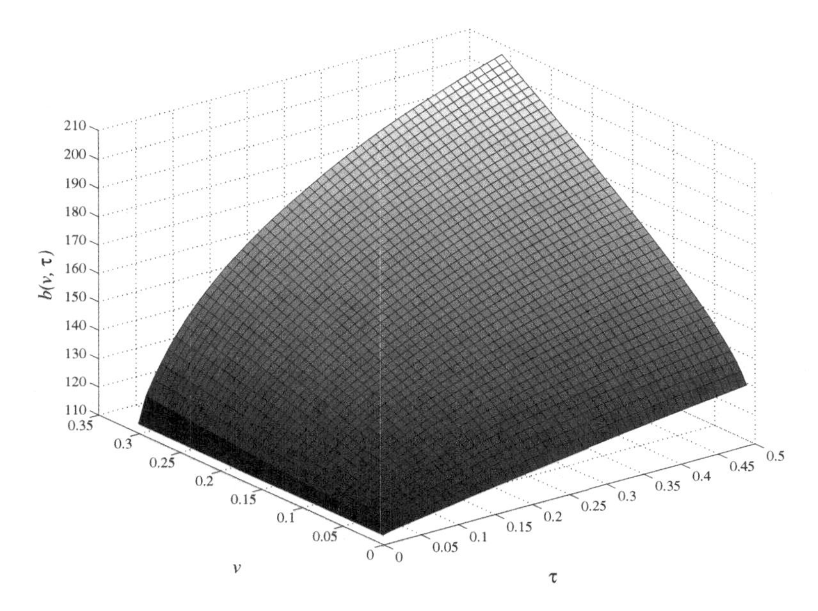

Fig. 7.1 Early exercise surface for a 6-month American call option, generated using the method of lines. Parameter values are as listed in Table 7.1 and $\rho = -0.5$.

at each iteration with respect to the integral term, and each spline is fitted using every grid point in S.

An alternative treatment of the jump integral for constant volatility is presented in Meyer (2015), where the mesh points of the Riccati transformation include the points required for a two point composite Gaussian quadrature of the jump integral. This approach eliminates the need for the spline interpolation in (7.12). However, we have no comparative data for its efficiency in the SVJD setting.

A sample early exercise surface is provided in Figure 7.1, generated using the method of lines for the case where $\rho = -0.5$. The value of the free boundary at expiry is independent of v. The free surface, $a(v, \tau)$, is an increasing function of v, and along a given value of v we observe an early exercise boundary of the form typically found for American call options. It is worth noting that the free surface generated by the method of lines is smooth, even when jumps are present, a feature not often displayed in the free boundary estimates generated using finite difference methods, such as Ikonen and Toivanen (2004).

We contrast the relative pricing accuracy of both the method of lines (MOL) and componentwise splitting (CS) methods. Using the parameter values in Table 7.1, and setting the spot volatility to $v = 0.04$, we generate American call option prices for S values of $80, 90, 100, 110$ and 120 using the three numerical methods discussed above; the method of lines as outlined in Sec. 7.4, the componentwise splitting approach which is detailed in Sec. 7.3, and a Crank-Nicolson scheme where the system of difference equations is solved using PSOR. The integral term is approximated in the same manner as for the MOL and CS, and we iterate at each time step, updating the integral term until the price profile converges.[5] We use $J = 50$ abscissa points for the integral term in all methods. We experimented with each of the above methods with an increase in the number of abscissa points from 50 to 100, but this led to a significant increase in the run time with no significant change in the option prices.

For both finite difference methods, we find that it is more efficient to update the integral term external to solving the system of difference equations.[6] Note that while we do not prove convergence for these iterative schemes with respect to value of the integral term, convergence is always observed in practice for the parameter values under consideration. The source code for all methods was implemented using NAG Fortran with the IMSL library running on the UTS, Faculty of Business F&E HPC Linux Cluster which consists of 8 nodes running Red Hat Enterprise Linux 4.0 (64bit) with 2×3 GHz 4MB Cache Xeon 5160 (dual core) Processors, 8 GB 667 MHz DDR2-RAM.

In the following, we treat the solution for the price from PSOR with 1,000 time steps, 3,000 volatility steps and 6,000 share price steps as the "true" solution for the purpose of comparing the efficiency of both the MOL and CS methods. We treat the delta and gamma from MOL with 500 time steps, $1,000$ volatility steps and $11,380$ share price steps as the "true" deltas and gammas. We compute the root mean square relative difference[7]

---

[5]Specifically, the system of difference equations is solved using PSOR for a given estimate of the integral term. The system of difference equations is then solved again using an updated estimate for the integral term. This is repeated until the price profile converges.

[6]For the componentwise splitting method, the five tridiagonal systems are solved sequentially for a given estimate of the integral term. The integral term is then updated, and the tridiagonal system solved again, repeating the procedure until the price profile converges.

[7]RMSRD is calculated as: $\sqrt{\frac{1}{5}\sum_{i=1}^{5}(\frac{\hat{C}(S_i)-C(S_i)}{C(S_i)})^2}$, where $S_i = 80 + 10 \cdot (i-1)$, $\hat{C}(S)$ is the estimate of the price, and $C(S)$ is the true price. It is important to use RMSRD to measure the errors from price, delta and gamma together since they have quite different numerical scales.

(RMSRD) using the option prices with S values of $80, 90, 100, 110$ and 120 with a spot volatility $v = 0.04$.

In Tables 7.4 and 7.5 we provide the American call prices produced by the method of lines (MOL), componentwise splitting (CS) and PSOR. As a basis for comparison, we also provide runtimes for the different methods, and compute the root mean-square relative differences (RMSRDs) for each method in relation to the "true" solution from PSOR. This allows us to make some further observations about the relative performance of the other two methods. We find that increasing the number of time steps for the MOL has little impact on the prices, while the price accuracy improves more when the grid size in the volatility and share price directions is refined. Clearly, the CS method runs faster when the grid is small compared with the MOL; with RMSRD as high as 0.7588% within 50 seconds it is almost 10 times faster than the MOL and produces results even more accurate than the MOL. However the accuracy deteriorates when the number of time steps is increased somewhat. It is also observed that the CS method produces prices with some oscillations especially in the negative correlation case.

If one wishes to achieve a given level of accuracy, we can see from the tables that the MOL seems the best method as it produces a higher accuracy with a RMSRD of 0.0177% within just half the time of CS, which produces results with a RMSRD of 0.0654%. Thus we are confident in asserting that the method of lines is very competitive for evaluating American options under stochastic volatility and jump-diffusion.

To see the overall efficiency of the three methods, we plot in Figures 7.2–7.5 below the comparisons of the average accuracy of the American call price, delta and gamma with the MOL, the CS method and PSOR. Note that these figures are plotted up to the point on the horizontal axis where the "exact" solution curves start to become artificially steep due to the fact that these are the solutions taken to be the "exact" ones. The RMSRD on the vertical axis in each of Figures 7.2–7.4 is calculated for the corresponding set of values, call prices, deltas or gammas, at the underlying prices $S = 80, 90, 100, 110$ and 120 and with both correlations $\rho = \pm 0.5$. In Figure 7.5 we display the average of the sum of the above three RMSRDs. The runtime on the horizontal axis in all graphs is the computer time taken to produce all three quantities, namely the call prices, deltas and gammas.

A number of comments based on the figures and the calculations are warranted. First, it is clear from Figure 7.5 that the method of lines performs best in calculating, to similar accuracy, the call prices, deltas and gammas. For instance, it costs MOL around 1,000 seconds to achieve an

Table 7.4 American call prices computed using method of lines (denoted MOL), componentwise splitting (denoted CS) and Crank-Nicolson with PSOR (denoted PSOR). Parameter values are given in Table 7.1, with $\rho = 0.50$ and $v = 0.04$. For the CS method, the first number in brackets for the CS method indicates the ratio between the grid step sizes at $S_{\max}$ and K imposed on the the non-uniform grid in S.

$\rho = 0.50, v = 0.04$			S			RMSRD	Runtime
Method (N, M, S_{pts})	80	90	100	110	120	(%)	(sec)
MOL (50,100,1138)	1.4844	3.7123	7.6982	13.6686	21.3645	0.0387	485
MOL (200, 100, 1138)	1.4847	3.7130	7.6993	13.6697	21.3654	0.0302	1,162
MOL (200, 250, 2995)	1.4848	3.7146	7.7018	13.6715	21.3657	0.0177	12,120
CS (2.5) (200, 100, 294)	1.4841	3.7070	7.6806	13.6387	21.3357	0.2006	60
CS (2.5) (300, 100, 294)	1.4747	3.6853	7.6442	13.5972	21.3029	0.6315	72
CS (2.5) (300, 200, 549)	1.4770	3.7027	7.6868	13.6563	21.3537	0.2820	175
CS (2.5) (1000, 1000, 2764)	1.4825	3.7120	7.6996	13.6690	21.3628	0.0654	12,985
PSOR (200, 200, 300)	1.4960	3.7415	7.7507	13.7300	21.4103	0.1920	1,080
PSOR (500, 500, 1000)	1.4861	3.7181	7.7086	13.6793	21.3707	0.0837	16,269
PSOR (1000, 3000, 6000)	1.4843	3.7145	7.7027	13.6722	21.3653	—	3,041,756

Table 7.5 American call prices computed using method of lines (denoted MOL), componentwise splitting (denoted CS) and Crank-Nicolson with PSOR (denoted PSOR). See Table 7.1 for the parameter values, with $\rho = -0.50$ and $v = 0.04$. For the CS method, the first number in brackets for the CS method indicates the ratio between the grid step sizes at $S_{\max}$ and K imposed on the the non-uniform grid in S. This is explained in Sec. 7.3.

$\rho = -0.50, v = 0.04$			S			RMSRD	Runtime
Method (N, M, S_{pts})	80	90	100	110	120	(%)	(sec)
MOL (50,100,1138)	1.1369	3.3512	7.5922	13.8786	21.7156	0.0578	485
MOL (200, 100, 1138)	1.1370	3.3518	7.5932	13.8798	21.7168	0.0542	1,159
MOL (200, 250, 2995)	1.1363	3.3530	7.5959	13.8827	21.7191	0.0193	12,122
CS (2.5) (200, 100, 294)	1.1368	3.3526	7.5950	13.8807	21.7162	0.0404	50
CS (2.5) (300, 100, 294)	1.1233	3.3199	7.5440	13.8309	21.6834	0.7588	65
CS (2.5) (300, 200, 549)	1.1298	3.3433	7.5855	13.8734	21.7120	0.2833	163
CS (2.5) (1000, 1000, 2764)	1.1336	3.3501	7.5940	13.8808	21.7174	0.0995	12,707
PSOR (200, 200, 300)	1.1651	3.4050	7.6510	13.9196	21.7358	0.4983	1,026
PSOR (500, 500, 1000)	1.1394	3.3594	7.6035	13.8875	21.7210	0.1660	16,979
PSOR (1000, 3000, 6000)	1.1359	3.3532	7.5970	13.8830	21.7186	—	3,115,836

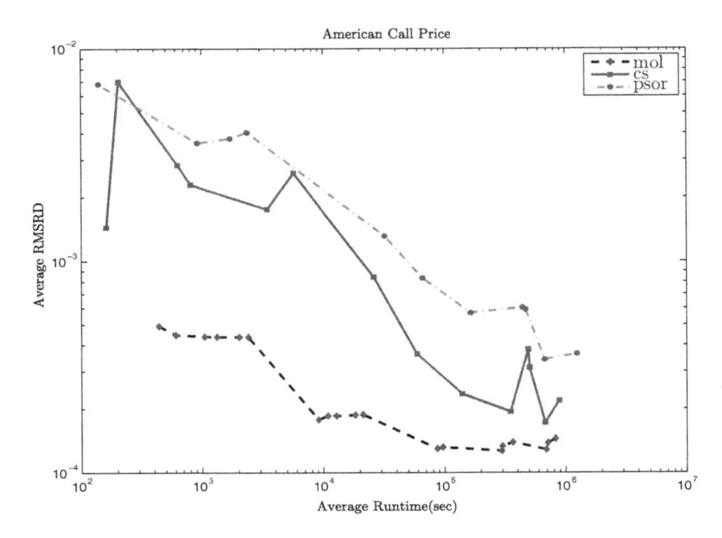

Fig. 7.2 Runtime efficiency of American call prices with MOL, CS and PSOR. We take the price from PSOR with a large grid consisting of 1,000 time steps, 3,000 volatility steps and 6,000 share price steps as the true solution. The root mean-square relative differences (RMSRDs) for each method in relation to this true solution correspond to the cases with share prices ranging from 80 to 120 and $\rho = \pm 0.5$.

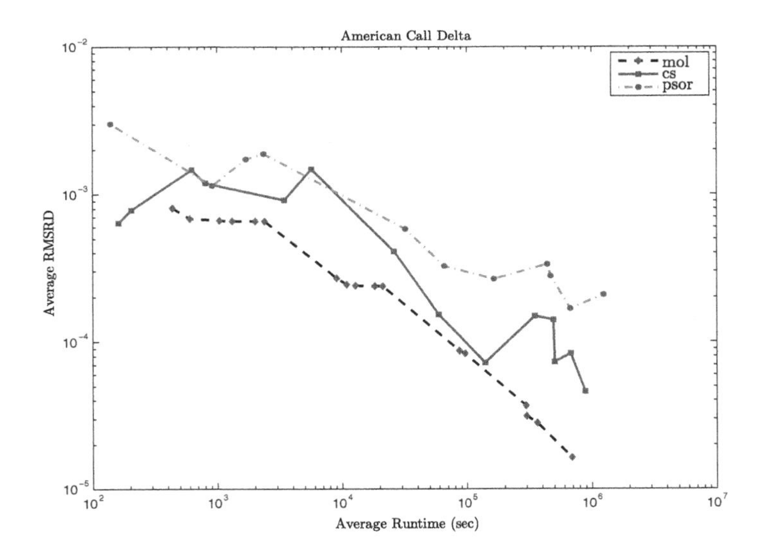

Fig. 7.3 Runtime efficiency of American call deltas with MOL, CS and PSOR. We take the delta from MOL with a large grid consisting of 500 time steps, 1,000 volatility steps and 11,380 share price steps as the true delta. The root mean-square relative differences (RMSRDs) for each method in relation to this true solution correspond to the cases with share prices ranging from 80 to 120 and $\rho = \pm 0.5$.

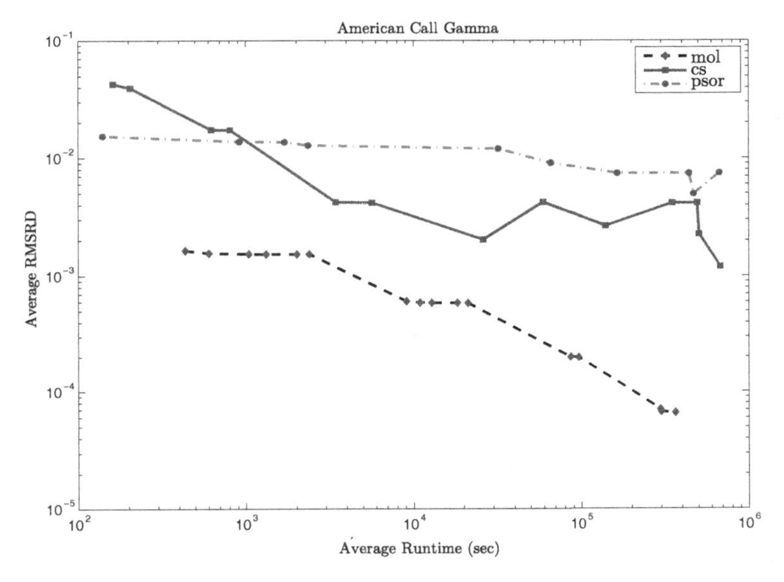

Fig. 7.4 Runtime efficiency of gammas with MOL, CS and PSOR. We take the American call gammas from MOL with a large grid consisting of 500 time steps, 1,000 volatility steps and 11,380 share price steps as the true gamma. The root mean-square relative differences (RMSRDs) for each method in relation to this true solution correspond to the cases with share prices ranging from 80 to 120 and $\rho = \pm 0.5$.

overall relative accuracy of 10^{-3} while it costs CS around 1,000,000 seconds to achieve the same overall accuracy and it takes PSOR even longer. Second, the results from the calculations (see Figures 7.3 and 7.4) show that the American call deltas and gammas seem to have a faster convergence rate than the American call prices with the MOL. This is natural because for MOL, within each iteration, the call prices are obtained after working out the deltas and gammas to the same degree of accuracy. We also see from the calculations that the value of the deltas and gammas do not change up to 5 decimals when refining the grid size which is also evidence indicating that the deltas and gammas with the MOL are much closer to the "true" deltas and gammas than those for both CS and PSOR.

In Chapter 6, we demonstrated that the Fourier cosine expansion approach (COS) is able to produce American option prices under both stochastic volaitlity and jump-diffusion dynamics by formula (6.100) and we have results in Table 6.2 which shows that the COS method is both accurate and efficient. However since the American option prices calculated from formula (6.100) are based on the computation of Bermudan option

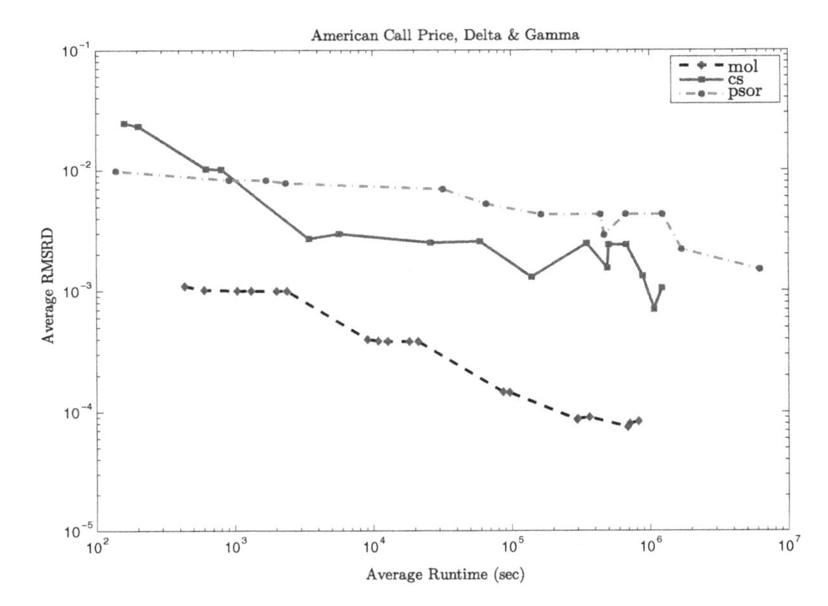

Fig. 7.5 Overall runtime efficiency with MOL, CS and PSOR. We take the solution from PSOR with a grid size of 1,000 time steps, 3,000 volatility steps and 6,000 share price steps as a true solution for the price. We take the delta and gamma from the MOL with a large grid consisting of 500 time steps, 1,000 volatility steps and 11,380 share price steps as the true delta and the true gamma. The root mean-square relative differences (RMSRDs) for each method in relation to this true solution correspond to the cases with share prices ranging from 80 to 120 and the correlations $\rho = \pm 0.5$.

with limited number of early exercise opportunities, hence we are able to obtain accurate and efficient option prices but it will take much longer to obtain a smooth early exercise boundary of the American option. Also the time taken to compute the option price by COS method is to compute only one option prices but the time taken by either the method of lines or finite difference method is to compute the option prices in a fine grid which carry some advantages. Hence, in summary, when the underlying is stochastic volatility and jump-diffusion dynamics, the COS method is efficient and relatively easy to implement to price either European option or Bermudan option with a number of early exercise opportunities. The meothd of lines is able to produce not only prices but also smooth and accurate early exercise boundaries of American options.

Chapter 8

Conclusion

This book has explored the pricing of American options. It has focused in particular on American call options but the techniques are generally applicable. We started in Chapter 2 with the case of the underlying asset dynamics following a jump-diffusion and stochastic volatility process.

In Chapter 3 we considered the case of jump-diffusion dynamics, in particular focusing on transform methods. We used the approach of Jamshidian (1992) to find an inhomogeneous partial-integro differential equation (PIDE) for the American call price in an unrestricted domain, which we then solved using Fourier and Laplace transforms, extending this very useful solution methodology to the jump-diffusion setting.

Chapter 4 considered American option pricing under stochastic volatility and jump-diffusion dynamics and also considered the transform approach. We saw how the transform techniques can be readily extended to the situation in which the volatility is stochastic.

Chapter 5 considered applying the transform methods to price American option under stochastic volatility only and we also applied the transform techniques to this situation in which only volatility is stochastic without jumps.

Chapter 6 considered American option pricing using the Fourier cosine expansion approach which is able to produce both European and Bermudan option prices accurately and efficiently and it is able to produce accurate American option prices with the help of the calculated prices of the Bermudan option with a certain number of early exercise opportunities.

Chapter 7 considered the more general case that allows for any general stochastic volatility and jump-diffusion terms. The chapter outlines three methods; the method of lines, the componentwise splitting approach of Ikonen and Toivanen, and the projected successive overrelaxation (PSOR) method that is used to generate the 'exact' solution. The chapter finds that the method of lines is the best method mainly because it calculates the delta and gamma of the option and the free boundary at the same time. The results are compared with those obtained using operator splitting methods and the Crank-Nicolson method with a high degree discretisation is used to obtain the 'exact' solution. The method of lines has been found to work quite well in the models considered here.

Other numerical methods have not been considered, such as the tree methods of Amin (1993) and Broadie and Yamamoto (2003), and various finite difference scheme implementations, including Andersen and Andreasen (2000) and d'Halluin *et al.* (2004).

Bibliography

Abramowitz, M. and Stegun, I. A. (1970). *Handbook of Mathematical Functions* (Dover, New York).

Achdou, Y. and Pirroneau, O. (2005). *Computational Methods for Option Pricing* (SIAM Frontiers in Applied Mathematics).

Adolfsson, T., Chiarella, C., Ziogas, A. and Ziveyi, J. (2013). Representation and Numerical Approximation of American Option Prices Under Heston Stochastic Volatility Dynamics, *Quantitative Finance Research Centre, University of Technology, Sydney,* Working Paper Series No. 327.

Amin, K. I. (1993). Jump Diffusion Option Valuation in Discrete Time, *Journal of Finance* **48**, 5: 1833–1863.

Amos, D. E. (1986). Algorithm 644: A portable package for Bessel functions of a complex argument and nonnegative order, *ACM Trans. Math. Software* **12**: 265–273.

Amos, D. E. (1985). A Subroutine Package for Bessel Functions of a Complex Argument and Nonnegative Order, Sandia National Laboratory Report SAND85-1018, Sandia National Laboratories, Albuquerque, NM.

Andersen, L. and Andreasen, J. (2000). Jump-Diffusion Processes: Volatility Smile Fitting and Numerical Methods for Option Pricing, *Review of Derivatives Research* **4**: 231–262.

Barles, G., Georgelin, C. A. and Jakobsen, E. R. (2013). On Neumann and Oblique Derivatives Boundary Conditions for Nonlocal Elliptic Equations, Tech. rep., available online at http://arxiv.org/pdf/1302.5568v3.pdf.

Barone-Adesi, G. (2005). The Saga of the American Put, *Journal of Banking and Finance* **29**: 2909–2918.

Barone-Adesi, G. and Whaley, R. E. (1987). Efficient Analytic Approximations of American Option Values, *Journal of Finance* **42**: 301–320.

Bates, D. S. (1996). Jumps and Stochastic Volatility: Exchange Rate Processes Implicit in Deutshe Mark Options, *Review of Financial Studies* **9**: 69–107.

Black, F. and Scholes, M. (1973). The Pricing of Corporate Liabilites, *Journal of Political Economy* **81**: 637–659.

Brennan, M. J. and Schwartz, E. S. (1978). Finite Difference Methods and Jump Processes Arising in the Pricing of Contingent Claims: A Synthesis, *The Journal of Financial and Quantitative Analysis* **13**, 3: 461–474.

Brennan, M. J. and Schwartz, E. S. (1977). The Valuation of American Put Options, *Journal of Finance* **32**: 449–462.

Briani, M., Chioma, C. L. and Natalini, R. (2004). Convergence of Numerical Schemes for Viscosity Solutions to Integro-Differential Degenerate Parabolic Problems arising in Financial Theory, *Numerische Mathematik* **98**, 4: 607–646.

Broadie, M. and Detemple, J. (1997). The Valuation of American Options on Multiple Assets, *Mathematical Finance* **7**, 3: 241–286.

Broadie, M., Detemple, J., Ghysels, E. and Torrès, O. (2000). American Options with Stochastic Dividends and Volatility: A Nonparametric Investigation, *Journal of Econometrics* **94**: 53–92.

Broadie, M. and Yamamoto, Y. (2003). Application of the Fast Gauss Transform to Option Pricing, *Management Science* **49**, 8: 1071–1088.

Carr, P. and Hirsa, A. (2003). Why be Backward? Forward Equations for American Options, *Risk* **16**, 1: 103–107.

Carr, P., Jarrow, R. and Myneni, R. (1992). Alternative Characterizations of American Put Options, *Mathematical Finance* **2**: 87–106.

Carr, P. and Madan, D. B. (1999). Option Valuation using the Fast Fourier Transform, *Journal of Computational Finance* **2**, 4: 61–73.

Chang, C.-C., Chung, S.-L. and Stapleton, R. (2007). Richardson extrapolation technique for pricing American-style options, *J. Futures Markets* **27**, 8: 791–817.

Cheang, G., Chiarella, C. and Ziogas, A. (2013). The Representation of American Options Prices under Stochastic Volatility and Jump-Diffusion Dynamics, *Quantitative Finance* **13**, 2: 241–253.

Chiarella, C., Kucera, A. and Ziogas, A. (2004). A Survey of the Integral Representation of American Option Prices, *Quantitative Finance Research Centre, University of Technology Sydney*, Research Paper No. 118.

Chiarella, C. and Ziogas, A. (2009). American Call Options under Jump-Diffusion Processes: A Fourier Transform Approach, *Applied Mathematical Finance* **16**, 1: 37–79.

Chiarella, C. and Ziogas, A. (2006). A Fourier Transform Analysis of the American Call Option on Assets driven by Jump-Diffusion Processes, *Quantitative Finance Research Centre*, Working Paper Series No. 174.

Chiarella, C. and Ziogas, A. (2005). Evaluation of American Strangles, *Journal of Economic Dynamics and Control* **29**: 31–62.

Clarke, N. and Parrott, K. (1999). Multigrid for American Option Pricing with Stochastic Volatility, *Applied Mathematical Finance* **6**: 177–195.

Cont, R. and Tankov, P. (2003). *Financial Modelling with Jump Processes* (Chapman and Hall/CRC, Boca Raton, Florida).

Cox, J. C., Ingersoll, J. and Ross, S. A. (1985). A Theory of the Term Structure of Interest Rates, *Econometrica* **53**, 2: 385–407.

Cox, J. C., Ross, S. A. and Rubinstein, M. (1979). Option Pricing: A Simplified Approach, *Journal of Financial Economics* **7**: 229–263.

Debnath, L. (1995). *Integral Transforms and their Applications* (CRC Press, Baton Rouge).

Detemple, J. and Tian, W. (2002). The Valuation of American Options for a Class of Diffusion Processes, *Management Science* **48**, 7: 917–937.

d'Halluin, Y., Forsyth, P. and Labahn, G. (2004). A penalty method for American options with jump diffusion processes, *Numerische Mathematik* **97**: 321–352.

Duffy, D. G. (2001). *Green's Functions with Applications* (Chapman and Hall/CRC, Boca Raton, Florida).

Duffy, D. J. (2006). *Finite Difference Methods in Financial Engineering: A Partial Differential Equation Approach* (Wiley; Har/Cdr edition).

Dufresne, D. (2001). The Integrated Square-Root Process, *Centre for Actuarial Studies, Department of Economics, University of Melbourne*, Research Paper No. 90.

Fang, F. and Oosterlee, C. (2011). A Fourier-Based Valuation Method for Bermudan and Barrier Options under Heston's Model, *SIAM Journal on Financial Mathematics* **2**: 439–463.

Fang, F. and Oosterlee, C. (2009). Pricing Early-Exercise and Discrete Barrier Options by Fourier-Cosine Series Expansions, *Numer. Math.* **114**, 1: 27–62.

Fang, F. and Oosterlee, C. (2008). A Novel Pricing Method for European Options Based on Fourier-Cosine Series Expansions, *SIAM Journal on Scientific Computing* **31**, 2: 826–848.

Feller, W. (1951). Two Singular Diffusion Problems, *The Annals of Mathematics* **54**: 173–182.

Fouque, J.-P., Papanicolaou, G. and Sircar, K. R. (2000). *Derivatives in Financial Markets with Stochastic Volatility* (Cambridge University Press, Cambridge).

Fouque, J.-P., Papanicolaou, G., Sircar, K. R. and Solna, K. (2003). Multiscale stochastic volatility asymptotics, *SIAM Journal of Multiscale Modeling and Simulation* **2**: 22–42.

Geske, R. and Johnson, H. E. (1984). The American Put Option Valued Analytically, *Journal of Finance* **39**: 1511–1524.

Glowinski, R., Dean, E. J., Guidoboni, G., Juárez, L. H. and Pan, T. W. (2008). Applications of Operator-Splitting Methods to the Direct Numerical Simulation of Particulate and Free-Surface Flows and to the Numerical Solution of the Two-Dimensional Elliptic Monge-Ampère Equation, *Japan Journal of Industrial and Applied Mathematics* **25**, 1: 1–63.

Gukhal, C. R. (2001). Analytical Valuation of American Options on Jump-Diffusion Processes, *Mathematical Finance* **11**: 97–115.

Heston, S. (1993). A Closed-Form Solution for Options with Stochastic Volatility with Applications to Bond and Currency Options, *Review of Financial Studies* **6**: 327–343.

Huang, J. and Pang, J. (1998). Option Pricing and Linear Comlementarity, *Journal of Computational Finance* **2**: 31–60.

Hull, J. and White, A. (1987). The Pricing of Options on Assets with Stochastic Volatility, *Journal of Finance* **42**: 281–300.

Ikonen, S. and Toivanen, J. (2007). Componentwise Splitting Methods for Pricing American Options under Stochastic Volatility, *International Journal of Theoretical and Applied Finance* **10**, 2: 331–361.

Ikonen, S. and Toivanen, J. (2004). Operator Splitting Methods for American Options with Stochastic Volatility, *Applied Mathematics Letters* **17**: 809–814.

Jacka, S. D. (1991a). Optimal Stopping and the American Put, *Mathematical Finance* **1**: 1–14.

Jacka, S. D. (1991b). Optimal Stopping and the American Put, *Mathematical Finance* **1**: 1–14.

Jamshidian, F. (1992). An Analysis of American Options, *Review of Futures Markets* **11**: 72–80.

Johnson, H. and Shanno, D. (1987). Option Pricing when the Variance is Changing, *Journal of Financial and Quantitative Analysis* **22**: 143–151.

Johnson, H. E. (1983). An Analytic Approximation for the American Put Price, *Journal of Financial and Quantitative Analysis* **18**: 141–148.

Ju, N. and Zhong, R. (1999). An approximate formula for pricing American options, *The Journal of Derivatives* Winter **7**, 2: 31–40.

Kahl, C. and Jäckel, P. (2006). Not-so-Complex Logarithms in the Heston Model, *Department of Mathematics, University of Wuppertal*, Working Paper.

Kallast, S. and Kivinukk, A. (2003). Pricing and Hedging American Options using Approximations by Kim Integral Equations, *European Finance Review* **7**: 361–383.

Karatzas, I. (1988). On the Pricing of American Options, *Applied Mathematics and Optimization* **17**: 37–60.

Kim, I. J. (1990). The Analytic Valuation of American Options, *Review of Financial Studies* **3**: 547–572.

Kou, S. G. (2002). A Jump-Diffusion Model for Option Pricing, *Management Science* **48**, 8: 1086–1101.

Lee, R. W. (2004). Option Pricing by Transform Methods: Extensions, Unification and Error Control, *Journal of Computational Finance* **7**, 3: 51–86.

Lewis, A. L. (2000). *Option Valuation under Stochastic Volatility* (Finance Press, California).

Lieberman, G. M. (1996). *Second Order Parabolic Differential Equations* (World Scientific Publishing Company, Singapore).

Longstaff, F. and Schwartz, E. (2001). Valuing American Options by Simulation: A Simple Least-Squares Approach, *Review of Financial Studies* **14**: 113–147.

Lord, R., Fang, F., Bervoets, F. and Oosterlee, C. (2008). A Fast and Accurate FFT-based Method for Pricing Early-Exercise Options under Lévy Processes, *SIAM Journal on Scientific Computing* **30**, 4: 1678–1705.

MacMillan, L. W. (1986). Analytic Approximation for the American Put Option, *Advances in Futures and Options Research* **1**: 119–139.

McKean, H. P. (1965). Appendix: A Free Boundary Value Problem for the Heat Equation Arising from a Problem in Mathematical Economics, *Industrial Management Review* **6**, 2: 32–39, Note: This is an Appendix to a paper by Samuelson, P. A. (1965). Rational Theory of Warrant Pricing, *Industrial Management Review* **6**: 13–31.

Merton, R. C. (1976). Option Pricing when Underlying Stock Returns are Discontinuous, *Journal of Financial Economics* **3**: 125–144.

Meyer, G. (2015). *The Time-Discrete Method of Lines for Options and Bonds: A PDE Approach* (World Scientific Publishing Company, Singapore).

Meyer, G. H. (1998). The Numerical Valuation of Options with Underlying Jumps, *Acta Mathematica* **47**: 69–82.

Meyer, G. H. and van der Hoek, J. (1997). The Evaluation of American Options with the Method of Lines, *Advances in Futures and Options Research* **9**: 265–285.

Myneni, R. (1992). The Pricing of the American Option, *The Annals of Applied Probability* **2**: 1–23.

Oleinik, O. and Radkevic, E. V. (1973). *Second-Order Equations with Non-Negative Characteristic Form* (Springer).

Oosterlee, C. W. (2003). On Multigrid for Linear Complementarity Problems with Application to American-Style Options, *Electronic Transactions on Numerical Analysis* **15**: 165–185.

Pagliarani, S., Pascucci, A. and Riga, C. (2011). Expansion formulae for local Lévy models, Tech. rep., available online at http://ssrn.com/abstract=1937149.

Parkinson, M. (1977). Option Pricing: The American Put, *Journal of Business* **50**: 21–36.

Peskir, G. (2005). On the American Option Problem, *Mathematical Finance* **15**, 1: 169–181.

Pham, H. (1997). Optimal Stopping, Free Boundary, and American Option in a Jump-Diffusion Model, *Applied Mathematics and Optimization* **35**: 145–164.

Reisinger, C. and Wittum, G. (2004). On Multigrid for Anisotropic Equations and Variational Inequalities: Pricing Multi-Dimensional European and American options, *Computing and Visualization in Science* **7**: 189–197.

Ruijter, M. J. and Oosterlee, C. W. (2012). Two-dimensional fourier cosine series expansion method for pricing financial options, *SIAM J. Sci. Comput* **34**, 5: B642–B647.

Runggaldier, W. (2003). Jump Diffusion Models, in S. T. Rachev (ed.), *Handbook of heavy tailed distributions in finance* (North-Holland: Elsevier): 169–209.

Samuelson, P. A. (1965). Rational Theory of Warrant Pricing, *Industrial Management Review* **6**: 13–31.

Scott, L. O. (1997). Pricing Stock Options in a Jump-Diffusion Model with Stochastic Volatility and Interest Rates: Applications of Fourier Inversion Methods, *Mathematical Finance* **7**, 4: 413–426.

Scott, L. O. (1987). Option Pricing when the Variance Changes Randomly: Theory, Estimation, and an Application, *Journal of Financial and Quantitative Analysis* **22**: 419–438.

Shampine, L. F. (1994). *Numerical Solution of Ordinary Differential Equations* (Chapman & Hall, New York).

Shephard, N. G. (1991). From Characteristic Function to Distribution Function: A Simple Framework for the Theory, *Econometric Theory* **7**: 519–529.

Stein, E. M. and Stein, J. C. (1991). Stock Price Distributions with Stochastic Volatility: An Analytical Approach, *Review of Financial Studies* **4**: 727–752.

Strang, G. (1968). On the construction and comparison of difference schemes, *SIAM Journal of Numerical Analysis* **5**: 506–517.

Touzi, N. (1999). American Options Exercise Boundary when the Volatility Changes Randomly, *Applied Mathematics and Optimization* **39**: 411–422.

Tzavalis, E. and Wang, S. (2003). Pricing American Options under Stochastic Volatility: A New Method using Chebyshev Polynomials to Approximate the Early Exercise Boundary, *Department of Economics, Queen Mary, University of London*, Working Paper No. 488.

Wiggins, J. B. (1987). Option Values under Stochastic Volatility: Theory and Empirical Estimates, *Journal of Financial Economics* **19**: 351–372.

Wilmott, P., Dewynne, J. and Howison, S. (1993). *Option Pricing: Mathematical Models and Computation* (Oxford Financial Press).

Yanenko, N. N. (1971). *The Method of Fractional Steps: The Solution of Problems of Mathematical Physics in Several Variables* (Springer-Verlag).

Yang, C., Jiang, L. and Bian, B. (2006). Free Boundary and American Options in a Jump-Diffusion Model, *European Journal of Applied Mathematics* **17**: 95–127.

Zhu, J. (2000). *Modular Pricing of Options: An Application of Fourier Analysis* (Springer-Verlag, Berlin).

Zvan, R., Forsyth, P. A. and Vetzal, K. R. (1998). Penalty Methods for American Options with Stochastic Volatility, *Journal of Computational and Applied Mathematics* **91**: 199–218.

Index

About the Authors

Carl Chiarella is Professor of Quantitative Finance at the University of Technology, Sydney where he teaches courses in advanced instruments, derivatives, synthetic finance products and financial decision making under uncertainty. His research interests include derivative securities pricing, term structure of interest rates, quantitative finance techniques, disequilibrium macroeconomics, asset pricing theory and empirics.

Boda Kang is a Lecturer in Mathematical Finance in the Department of Mathematics at the University of York in the United Kingdom. Before joining York, he was a Senior Research Associate in the Finance Discipline Group and the Quantitative Finance Research Center at the University of Technology, Sydney for about 6 years. Before that he was completing his PhD under the direction of Professor Jerzy Filar. His research interests include financial derivatives pricing, computational finance, financial mathematics, energy and volatility derivatives modeling, time-consistent dynamic risk measures, Markov decision processes and their applications.

Gunter Meyer is Professor Emeritus of Mathematics at the Georgia Institute of Technology in Atlanta, where he helped develop and taught in the MS program in quantitative and computational finance. His research interests include numerical methods for partial differential equations, free boundary problems, reaction diffusion problems in finance, reaction diffusion problems and numerical heat transfer, hysteresis, two-point boundary value problems for ordinary differential equations and hydrodynamic stability.

Printed in the United States
By Bookmasters